EMC for Installers

EMC for Installers
Electromagnetic Compatibility of Systems and Installations

Mark van Helvoort and
Mathieu Melenhorst

CRC Press
Taylor & Francis Group
Boca Raton   London   New York

CRC Press is an imprint of the
Taylor & Francis Group, an **informa** business

CRC Press
Taylor & Francis Group
6000 Broken Sound Parkway NW, Suite 300
Boca Raton, FL 33487-2742

First issued in paperback 2020

ISBN-13: 978-1-4987-0248-5 (hbk)
ISBN-13: 978-0-367-65705-5 (pbk)

Library of Congress Cataloging-in-Publication Data

Names: Helvoort, Mark van, author. | Melenhorst, Mathieu, author.
Title: EMC for installers: electromagnetic compatibility of systems and installations / Mark van Helvoort and Mathieu Melenhorst.
Description: Boca Raton, FL: CRC Press/Taylor & Francis Group, 2018. | Includes bibliographical references.
Identifiers: LCCN 2018027170| ISBN 9781498702485 (hardback: acid-free paper) | ISBN 9781315119359 (ebook)
Subjects: LCSH: Electromagnetic compatibility. | Shielding (Electricity) | Electric apparatus and appliances—Installation.
Classification: LCC TK7867.2 .H45 2018 | DDC 621.382/24—dc23
LC record available at https://lccn.loc.gov/2018027170

Visit the Taylor & Francis Web site at
http://www.taylorandfrancis.com

and the CRC Press Web site at
http://www.crcpress.com

In memory of Prof. Dr.Ir. P.C.T. van van der Laan.

Contents

Preface

At the beginning of our careers, almost three decades ago, electronics was not yet all abundant in our society, and electromagnetic compatibility (EMC) was practiced by a small number of specialists, mainly the domain of the military (in the United States and Russia) or of high-voltage engineering (in Europe) where electromagnetic disturbances were more severe than elsewhere. Nowadays, EMC has become embedded in the development process of all equipment, partially born by necessity, partially forced upon by legislation put in place all around the globe.

Many books and guidelines have been written over this period, but surprisingly few provide practical information on EMC planning and design of large installations and complex systems. In 2011, we published a successful hands-on book in Dutch for installers: *EMC of Installations—A Path towards Electromagnetic Compatibility.* Based on its success, we decided to write this more extended and international book.

In this book, we will discuss those EMC aspects that are relevant for installations and large systems. A strong focus is placed on proper grounding concepts, cabling and wiring, the protection of cabling, zoning, and shielding between zones. Explanations and templates are provided for EMI risk inventory (source–victim and zone compatibility matrix). Pre- and full-compliance testing is described, and the final, short chapter involves a structured approach to troubleshooting.

Acknowledgments

Mathieu Melenhorst: I express my profound gratitude to my late parents: Jan Melenhorst and Henny Alsters. My brother Frans Melenhorst for everything he has done for me, his wise words and in admiration for his strength. I thank Frits Buesink of the Twente University of Technology and Thales for planting the EMC seed when I was a trainee and being a wonderful colleague at Thales. I also thank Corné van Sommeren for giving me the opportunity to start the actual EMC craftsmanship, when he was my manager at FEI/Philips Electron Optics, now Thermo Fisher Scientific. Last but certainly not least: my wife Klaartje Hermans for her patience and support without which this never could have been possible.

 Mark van Helvoort: I express my gratitude to my parents who stimulated me in my scientific and technical career. I thank Bert van Heesch who sparked my interest in high-voltage engineering, which in turn led to my PhD in electromagnetic compatibly, under highly valued guidance of Lex van Deursen, nearly 25 years ago. In addition, I also thank my former and current colleagues at Eindhoven University of Technology, AMP and Philips, for their interesting discussions and their willingness in supporting this book with ideas and materials. In particular, I want to mention Dick Harberts; I greatly appreciate our joint writing of scientific papers. Further, I acknowledge the support received from the European Eureka CATRENE and PENTA programs for funding the THOR and DISPERSE projects that paved the way for this book. Here I need to mention Paul Merkus who taught me the tricks of trade in public–private partnerships and Ernst Hermens for his excellent support in these projects. Most importantly, I am highly indebted to my loving and supportive wife, Debby, and my three gifted children Pim, Tom, and Bram for everything beyond technology.

Both authors express their gratitude to the companies and their employees who were helpful in retrieving information or images needed for this book: Aline Smelt of Hemmink (Pflitsch), Bernhard Mund of Bedea, Toine van Esch and Mattie de Leest of Philips MRI, Wilko de Graaf, Khadir Hadfoune and Toon van Til of Philips QLAB EMC Laboratory, Rob Kleihorst of Philips IGT, Philips Refurbishment, Jan van Schie of TKF, and Guus Pemen from Eindhoven University of Technology.

Authors

Dr. Ir. Mark van Helvoort, PMP, PBA, is a program manager at Philips responsible for public–private partnerships related to magnetic resonance imaging (MRI). MRI scanners are complex systems with an inherently challenging electromagnetic environment. As project manager, he is currently involved in investigating the coexistence of active medical implantable devices and MRI. Before this function, he was a hardware architect responsible for system electromagnetic compatibility (EMC) design and a group leader responsible for RF electronics. Before joining Philips in 1999, he was with AMP (currently TE Connectivity) where he was responsible for very high-speed connector simulations and investigations related to automotive wire harnessing and EMC aspects of premises networks. He was responsible for the installation of an EMC test lab for automotive components. He received his PhD from Eindhoven University in 1995. His thesis described the grounding structures for the EMC protection of cabling and wiring. Parts of his thesis were adapted for publication in a Dutch journal for installers (*Installatie Journaal*). In 2011, van Helvoort and Melenhorst authored a Dutch book on EMC *EMC van Installaties—Op weg naar eletromagnetische compatibiliteit*. In total, van Helvoort, at the time of writing, holds 77 publications and 29 patents. He is a senior member of IEEE.

Mathieu Melenhorst, BSc, is a seasoned building service consultant specialized in electromagnetic phenomena at Sweco, the Netherlands, with a track record in electromagnetic compatibility of large installations and complex systems. In his current role, he is responsible in a wide variety of projects for preventing electromagnetic interaction between equipment and between equipment and humans (electromagnetic compatibility [EMC] and electromagnetic field), electrostatic discharge (ESD), lightning protection, stray current management (corrosion prevention), and power quality. Before joining Sweco, Melenhorst worked at Alewijnse on marine systems, at Croon Elektrotechniek, on infrastructural projects as a specialist in EMC and lightning. He was involved as an EMC engineer in radar and optical systems for naval vessels at Thales Nederland and as an EMC designer for electron microscopy systems at FEI/Philips Electron Optics (currently Thermo Fisher Scientific). Melenhorst has written a good number of technical papers, including a book on EMC in Dutch, together with Mark van Helvoort. He is member of the Dutch ESD–EMC society and a gifted speaker at seminars.

CHAPTER ONE

Planning for electromagnetic compliance

Complex systems and large installations should operate reliably without interruption or degraded performance due to electromagnetic interference. Enforced by laws around the globe, all (electronic) equipment has to fulfill certain requirements on the emission of, and the immunity to, electromagnetic disturbances. Compliance to these laws enables harmonic coexistence of equipment. In technical terms, this compliance is called electromagnetic compatibility, which is better known by its acronym EMC.

The International Electrotechnical Commission (IEC 60050-161 2017) defines EMC as

> The ability of equipment or a system to function satisfactorily in its electromagnetic environment without introducing intolerable electromagnetic disturbances to anything in that environment

Obviously, the term "disturbance" is key in this IEC definition. First, the classic source–victim model will be discussed, including the source–victim matrix analysis tool for risk assessment of individual equipment. Next, the important, yet complex, concept called "environment" will be introduced. From this, the zoning concept will be derived, which in turn can be managed with a zone compatibility matrix. Both the source–victim matrix and the zone compatibility matrix are used as risk assessment tools to identify where specific EMC mitigations have to be designed. Basic EMC guidelines for systems and installations, including mitigation measures to EMC risks, are discussed in the subsequent chapters.

1.1 Source–victim model

Intolerable electromagnetic interference requires both a *source* for creating a disturbance and a *victim* for which this disturbance is intolerable. In addition, a *path* for the disturbance to travel from the source to the victim is essential. This classic EMC model is depicted in Figure 1.1.

In this model, the source may be either natural or man-made (Degauque 1993). Examples of *natural noise* are thunderstorm activity, lightning strokes, and magnetic storms. *Man-made noise* could be arc welding, harmonics from variable frequency drives, or RF leakage from a microwave oven. Both natural noise and man-made noise are called *emission* by IEC: Any EM emission, natural or "man-made", is potentially a disturbance to any other susceptible device in the environment. *Susceptibility* is defined as the threshold at which other equipment will start to malfunction or break down due to an external electromagnetic disturbance. *Immunity* is the opposite, it being the ability of equipment to function correctly in the presence of electromagnetic interference. *Hardening* is the discipline to reduce the susceptibility or increase the immunity of equipment. Immunity should not be mistaken for invulnerability: any electronic system will break down above a certain threshold.

Disturbances may travel over cables (conducted) or through the air as electric (parasitic capacitance) or magnetic fields (parasitic mutual inductance) or as electromagnetic waves

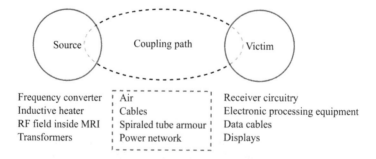

FIGURE 1.1 The basic EMC model contains three essential elements: the source, the victim, and a path for the disturbance to travel from the source to the victim.

(radiated). This is called the *coupling path*. Without coupling path, the disturbance from the source will not be able to interact with the victim, and the victim will remain to function unimpaired.

1.1.1 Source–victim matrix

A source–victim matrix is created to evaluate the risk of interference between equipment parts. A sample source–victim matrix is shown in Figure 1.2. This matrix lists potential sources of disturbance as rows and the victims, for which this disturbance is intolerable, as columns.

A victim is sometimes referred to as "sink." As will become obvious from the subsequent chapters, this term may lead to confusion. For any combination of source and victim, the risk of interference has to be assessed and entered at the crossing of the row of the source and the column of the victim in the matrix. When interference is possible, a "1" is entered; if it is likely, a "2" is entered. When interference is very unlikely, the crossing is left blank. Any equipment can be both a source for disturbance and simultaneously a victim for another disturbance, so the full matrix has to be filled.

When the risk assessment shows that interference is possible, mitigation measures have to be described. When interference is likely, avoidance measures must be analyzed and described. An example is shown in Table 1.1. In this case, the frequency drive is designated as the source and the safety-PLC as the victim, and a high likelihood of interference is predicted. This is indicated with "2" at coordinate A2.

Date	30-3-2018
Page number	1
Number of pages	1
Object	Tunnel equipment room
Author	M. Melenhorst

Legend	
	No interference likely
1	Interference possible
2	Interference likely
3	Other (add description)

Source \ Victim (sink)	1 Frequency drive	2 Safety PLC	3 Motor control cabinet (MCC)	4 Hand held transmitter	5 C2000 wireless emergency communications system	6	7	8	9	10	11	12	13	14
A Frequency drive		2	2	2	1									
B Safety-PLC			1	1										
C MCC														
D Hand held transmitter	1	1												
E C2000 wireless emergency communications system	1			1										
F														
G														

FIGURE 1.2 Example of the source–victim matrix.

Table 1.1 Overview of mitigation and avoidance measures based on the risk assessment noted in the source–victim matrix

Row	Column	Description	Mitigation and avoidance
A	2	Risk of interference due to high level of conducted and radiated noise of the frequency drive	Frequency drive selection, shaft brushes, earthing and bonding, cable selection of frequency drive and use of trays as PEC. Put safety PLC in EMC enclosure
A	3	Interference of the analog circuitry of the motor control cabinet (MCC)	As A2, cable selection, classification, and spacing of MCC cables. Use cable trays or put analog cable in pipe. If possible and allowed (delay): filtering of analog signals
A	4	High-pitch noise during transmission and reception	Shielding and filtering at the frequency drive. Cabling, earthing, and bonding
A	5	Transmission and reception outages, lowered quality of service	Shielding and filtering at frequency drive (as above). Note that C2000 has top priority (safety aspects)
B	3	Interference of high-impedance analog signals	Choose other analog cable, reroute or use piping, safety PLC in EMC enclosure, filtering of analog signals
B	4	Audible noise during Rx	Increase distance
D	2	Safety PLC mishaps	Use EMC enclosure to shield safety PLC. If possible, add warning sign to keep distance when using handhelds
D	3	MCC operation interference	Add warning sign to keep the distance between handhelds and the MCC
E	2	Safety PLC interfered	Use EMC enclosure for safety PLC, cabling and bonding practices, safety PLCs and C2000 shall be made compatible
E	4	Audible noise on handhelds	C2000 has priority over handheld transmitters, add notification in user's instructions

1.2 Environments

All equipment is exposed to a variety of phenomena. For example, the camera in Figure 1.3 can be shaken, sometimes dropped and used, or stored at different temperatures. These phenomena can be clustered according to their similarities in effect, for example, the first two involve shock. Such a cluster is called an *environment*. Every system or installation is subjected to a combination of environments. In a dynamo, a rotating magnet generates an electric voltage on purpose. In case the motion of a magnet is not intentional, the effect may be an electromagnetic

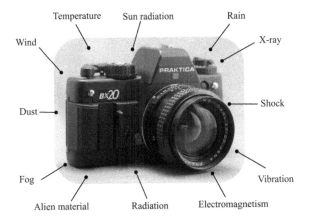

FIGURE 1.3 The photo camera, its environments, and the environmental effects.

disturbance. It is for this reason that the interaction between the electromagnetic environment and the installation focuses on the susceptibility of the installation.

In complex systems and large installations, the size of the source–victim will grow to an extent that it becomes impractical to manage. Consider, for example, a (naval) vessel in which a multitude of systems have to be integrated in a limited space. It is therefore necessary to focus on the most important sources and victims.

1.2.1 **Location classes**

In many situations, the electromagnetic environment will be predefined by natural noise and man-made noise originating from an already installed equipment. The IEC defines three typical or archetypal location classes and related electromagnetic environments based on the prevailing electromagnetic phenomena (IEC 61000-2-5, 2017):

1. Residential: The residential location exists in an area of land designated for the construction of domestic dwellings, which is a place for one or more people to live.

2. Commercial: Commercial/public location is defined as the environment in areas of the center of a city, offices, public transport systems (road/train/underground), and modern business centers containing a concentration of office automation equipment.

3. Industrial: Industrial installations are characterized by the fact that many items of equipment are installed together and operated simultaneously, and some of these items of equipment might act as a severe interference source.

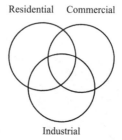

Residential　　Commercial

Industrial

FIGURE 1.4　Graphical representation of location classes in EMC.

These locations are graphically depicted in Figure 1.4. When the location classes do not overlap, the environment class corresponds to the location.

1.2.2 **User and intended environment**

The question if a product will operate satisfactorily when immersed in a certain electromagnetic environment implicitly addresses two important aspects:

1. The intended environment: This is the environment which the manufacturer of the equipment has designed their product for.
2. The user environment: This is the environment in which the equipment will be installed and used "for real."

The aspects show that the manufacturer can define the intended environment, while the user or system integrator often has to cope with the user environment. If, for a certain system, the intended environment and user environment coincide, the system should be sufficiently immune to interference in the user environment. On the contrary, it will not add an unacceptable interference to its environment in order to realize the interference-free operation of other equipment.

Great care must be taken to verify that the intended and user environments really coincide. The manufacturer often poses additional conditions in order to establish compliance with certain environments, for example, the type and length of cabling or the need for filtering.

Both U.S. Federal Communications Commission (FCC) and European Union directives relate the electromagnetic compliance of equipment to the intended environment. Typically, this information can be found in the *declaration of conformity,* which may also list or refer to the conditions under which compliance is met.

1.3 Zones and barriers

In specific situations, for sensitive equipment like an electron microscope, which is susceptible to mechanical vibrations, draught, and very-low-frequency magnetic fields, the manufacturer will pose strict constraints on the environment in which the microscope has to operate in order to achieve its specifications. It will be too large a constraint for an installation to subject all equipment to this strict specification. Therefore, a large installation or system is split in multiple electromagnetic environments called *zones*.

The zone concept becomes a powerful tool for risk assessment when measurable and testable attributes within the EMC domain are assigned to each zone. This zone can be interpreted as a local user environment. The EMC attributes of the installed equipment are provided by the manufacturer. The risk of electromagnetic interference (EMI) is low when the attributes of the installed equipment and its zone coincide. Mitigation measures such as defining a new zone or hardening can be used if the risk of EMI is deemed too high. Another possibility is *avoidance* of the risk by selecting an alternative equipment which is fit for purpose in this zone.

Systems within a single zone shall be electromagnetic compatible with each other. This is the whole purpose and idea behind zoning.

As shown in Figure 1.4, sometimes location classes overlap. An industrial installation, for example a car wash installation, can be installed in a shopping mall, which itself is a commercial environment. Another example is welding equipment used in a residential area. The welding equipment may cause voltage dips on the mains network, which are compliant with the industrial environment, but surpass the limits for household appliances. The electromagnetic phenomena in overlapping location classes blend, and zoning might be appropriate to enable mitigation measures.

1.3.1 Barriers

Two zones are separated from each other by a *barrier*. A barrier blocks electromagnetic interference traveling from one zone to another (IEC/TR 61000-5-6, 2002). Without barrier, the equipment designed for different environments will interfere with each other. Chapter 6 will discuss practical implementations of barriers such as *filters*.

It is convenient and useful to use the physical boundaries in a large system or installation as demarcation for the enclosed electromagnetic environment. A zone is therefore a volume, limited by physical boundaries, and described by its electromagnetic phenomena. Often, but not always, these physical boundaries act as barrier themselves or can be redesigned as such (see Chapter 7 for shielding fields).

1.3.2 Zone assignment

Each zone provides a level of protection where the number increases from the outside to the inside, like the rings in a tree. Each tree ring forms a barrier that provides a level of protection to the volume or zone behind it (see Figure 1.5). An example is given in Figure 1.6, where Zone 1 is, for example, an industrial environment described by IEC (IEC 61000-6-2, 2005), Zone 2 is a residential environment described by IEC (IEC 61000-6-1, 2005), and Zone 3 is a *protected environment*, which means a custom-defined electromagnetic environment with little disturbances.

FIGURE 1.5   Tree rings representing zones.

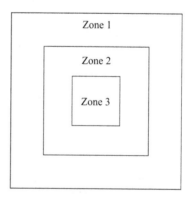

FIGURE 1.6   Numbered zones.

Obviously, the amount of electromagnetic disturbance that the equipment is allowed to add to a zone is limited. The emission levels are described by IEC:

- Zone 1, industrial environment: IEC 61000-6-4, 2018
- Zone 2, residential, commercial, and light-industrial environment: IEC 61000-6-3, 2010

Since residential, commercial, and light industrial environments might be co-located (like a car wash in shopping mall with residential apartments above them), the emission categories for these categories have been clustered in a single standard. The emission limits in IEC 61000-6-3 are lower than those in IEC 61000-6-4. This means that, from an emission point of view, residential and commercial equipment can be used in an industrial environment, but not vice versa.

It is common practice to number the zones from the outside to the inside, where 0 is often used for the outside world. Guidelines to characterize the different zones are provided by IEC/TR 61000-2-5, covering the following phenomena:

- Low frequency
    - Harmonics and interharmonics
    - Voltage dips, variations, and interruptions
    - Voltage unbalance and frequency variations
    - Common mode voltages
    - Signaling voltages
    - Induced low-frequency voltages
    - DC in AC networks
    - Low-frequency magnetic fields
        - DC
        - Railway
        - Power system
        - Power system harmonics
        - Non-power systems-related fields
    - Low-frequency electric fields
        - DC fields
        - Railway
        - Power systems
- High frequency
    - Signaling voltage

- Power line transmission
- Conducted continuous wave
- Induced conducted continuous wave
- Transients
- Oscillatory transients
- Radiated
  - Pulsed
  - Modulated
- Electrostatic discharge

1.3.3 Zone compatibility matrix

The *zone compatibility matrix* is used to summarize the information about the user environment and the intended environment in order to detect possible mismatches which lead to risk of noncompliance. Consider, for example, an elevator which has been designed for a residential area (Zone 2 in Figure 1.5) but may be installed in an industrial environment (Zone 1).

To create the zone compatibility matrix, the following steps have to be taken:

1. Define the areas (locations) in a building or within an enclosure and assign the user environment to each area. An example is given in Figure 1.7.
2. Project the newly to-be-installed equipment on the building or within the enclosure. An example is given in Figure 1.8.

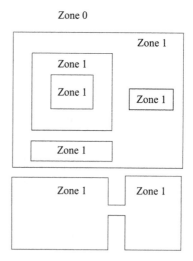

FIGURE 1.7   Mapping the user environment on locations or areas in a building or enclosure.

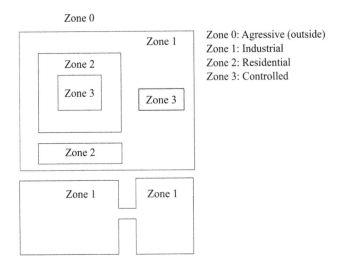

FIGURE 1.8 Projection of the intended environment for a newly to-be-installed equipment.

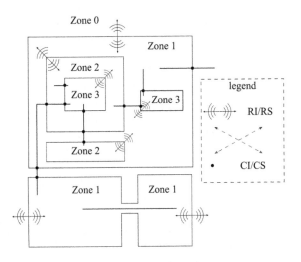

FIGURE 1.9 Potential coupling paths between the projected intended environments. RI/RS is radiated immunity/susceptibility and CI/CS is conducted immunity/susceptibility.

3. Visualize the parts that may compromise the effectiveness of zones defined in Figure 1.8 by mapping coupling paths: route the cables and identify the potential radiated effects. An example is given in Figure 1.9.

The building used in the example of Figures 1.7–1.9 consists of an exterior steel framework and an interior made from timber.

Table 1.2 The zone compatibility matrix compares the intended environment of equipment with the user environment

Component or environment	Intended environment	User environment	Mitigation and avoidance
Lift	3	1	A
Escalator	3	1	A
IT room	2	1	B
Telecom room	2	1	C
Radio room	3	1	D
Slide doors	3	1	E
Locker room	1	1	

Therefore, little or no inherent shielding between the zones can be expected and the user environment contains only two zones: Zone 0 for the outside of the building and Zone 1 for all interior areas. When the newly to-be-installed equipment is mapped on the building, the need for four zones is found (see Figure 1.8). The result can be summarized in the zone compatibility matrix as shown in Table 1.2, and the risks for incompatibility can be identified.

Before mitigations for these risks can be defined, the effectiveness of each zone has to be considered. For this purpose, all potential coupling paths between zones, both conducted and radiated, have to be considered. The coupling paths are mapped on the intended environment (Figure 1.9). Most often, cables are the primary radiators, thereby making it sometimes difficult to clearly separate both effects. A large part of this book (Chapters 2–4) will therefore focus on cabling and wiring.

After the coupling paths are added to the zone compatibility matrix, mitigations for each incompatibility between the intended and user environments have to be designed and entered in the "Mitigation/Avoidance" column of Table 1.2. With reference to Figure 1.9, examples for such measures are as follows:

A. Select a different lift and environment, which are suitable for the existing user environment (avoidance).

B. Use SFTP equipment and cabling (mitigation).

C. Put up signs that mobile communication equipment is forbidden in this zone (avoidance, risk reduction of occurrence).

D. Zone 3 in the example contains a radio room. Mitigation D proposes to shield this room and to add filters for power and signal cabling and EMC glands for communication cables at its entrance (mitigation).

E. Place the control electronics of the sliding doors in a shielded enclosure (mitigation).

The zone difference also gives an indication for the *attenuation* required from the *barrier* between the zones. If the available attenuation provided by the building structure is too low, additional shielding has to be applied as in the case of the radio room (see Chapters 5 and 6 for details on shielding and cable penetration). When adjacent zones only have a difference of one level, like the telecom room (Zone 2) next to an industrial environment (Zone 1), typically less attenuation is required than when the difference is two or more. The attenuation of the shielding box for the sliding door electronics, which provides a barrier between Zone 1 and Zone 3, will have to be higher than that for the barrier around the telecom room.

Since EMC has been defined as the ability of a piece of equipment to function satisfactorily in its environment without introducing intolerable electromagnetic disturbance to that environment, the introduction of equipment intended for a zone with a lower class than the user zone will likely result in degradation of the user zone to the class of the intended zone, with potential effects on the already installed equipment. For example, consider a Zone 2 environment in which a large UPS (uninterruptable power supply) is installed, which has been designed for Zone 1. In this situation, Zone 2 is degraded in to Zone 1, unless countermeasures are taken and a new zone around the UPS is defined.

1.3.4 **Safety**

Electromagnetic interference may also disturb safety-related functions. For any safety-instrumented function, great care must be taken when comparing the intended environment and the user environment. Depending on the safety integrity level (IEC 61508, 2010), more strict requirements are placed on the susceptibility of the equipment than may be expected from the zone classification. Therefore, the selected equipment needs to have a larger margin with respect to its susceptibility levels or needs an additional protection against the environment, which means the introduction of a new, dedicated zone (Zone 3).

1.4 Zone compatibility matrix versus source–victim matrix

Two matrices are described that appear to be similar (compare Figure 1.2 with Table 1.2). The zone compatibility matrix is used in case where the equipment is placed in the natural or induced environment. In this case, the effects of a source can be measured throughout that zone, and mitigation measures are difficult or impossible to take within the zone (without adding a new zone). A transformer, for example, can cause a high level of magnetic field in its surroundings. Shielding the transformer may prove impossible or impractical (though some counter-measures will be discussed in Chapter 7). In some situations, the electromagnetic environment can prove to have too large disturbances for a system to operate in.

The source–victim matrix, on the other hand, is defined by sources and victims that can be identified one on one. The sources and victims are placed in an environment where the background noise level is lower than the level they emit (source) or their susceptibility threshold (victim). This means that the man-made noise, generated by the equipment noise, dominates the present electromagnetic environment or zone. This also means that the source and the victim are operating within the same zone.

CHAPTER TWO

Grounding and earthing

The terms *"grounding"* and *"earthing"* are synonyms. The first term is common in the United States and the latter in the United Kingdom. Literally, grounding and earthing mean making a connection with ground and earth, respectively. In both cases, planet Earth is meant. Such a connection plays a vital part in lightning protection and power distribution. Electrotechnically, however, this term is used to identify all signal and low power return path even if they do not have a direct connection to planet Earth. This use stems from the first large installation: the telegraph. Telegraph wires were suspended on wooden poles between transmitter and receiver. To save copper the planet Earth (ground) was used as return path for the telegraph signals and grounding became a synonym for return path. In Electromagnetic Compatibility (EMC) context, the definition of grounding has to be extended to: "Ground is a low impedance path for current to the source" (Ott 1983).

This chapter discusses grounding in low-voltage power distribution networks, lightning protection and ground as signal return.

2.1 Power distribution

Most residential user and small to medium enterprises are connected to the low-voltage power distribution grid. The system grounding in these grids is designed for the safety of persons, livestock, and property and not necessarily for EMC. There are several types of system grounding, which are designated by two letters (IEC 60364-1 2005):

The first letter describes the system ground, which is the relationship between the power system and planet Earth:

- T: Direct connection to the planet Earth;
- I: All live parts isolated from planet Earth, or one point connected through a high impedance;

    The second letter describes the chassis ground, which is the relationship between a conductive chassis and planet Earth:

    - T: Direct electrical connection to planet Earth, independently of the grounding of the power system;
    - N: Direct electrical connection to the ground point of the power system.

The U.S. National Electric Code (NPFA 70) labels chassis ground as *equipment ground* (EG). The IEC calls chassis ground *protective earth* (PE) or protective ground. The first term is common outside the United States. The most common types of grounding systems are TT, TN, and IT. We will discuss them in more detail in the remainder of this section and establish the relationship between low-voltage power distribution grounding, lightning protection, and EMC.

2.1.1 **TT grounding system**

Figure 2.1 depicts a three-phase low-voltage distribution setup implementing the TT grounding system. The power source on the left hand side has a direct connection to planet Earth via the grounding rod label with $R_{r,s}$. The conductive equipment

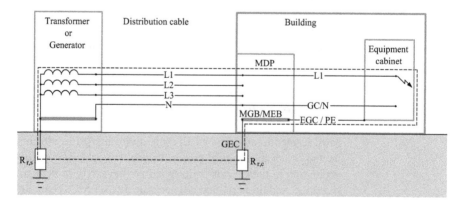

FIGURE 2.1 In TT systems a short circuit between phase and ground will lead to current flow (dashed line) through planet earth.

cabinet, which is powered by the power source, has its own direct connection to planet Earth via the grounding rod with label $R_{r,c}$. In US terminology the *Equipment Ground Conductor* (EGC) connects the cabinet to the Mains Ground Bar (MGB) in the *Mains Distribution Panel* (MDP). The mains ground bar in its turn is connected to the grounding rod with the *ground electrode conductor* (GEC).

In UK terminology the Mains Ground Bar is called *Mains Earth Bar* and the EGC is labeled with *PE*. The power source can either be the secondary of a transformer on the power grid, or a stand-alone generator (including fuel-fired, solar cells and wind mills).

In case of a short circuit between one of the phases and the equipment cabinet the current from this phase will return to the power source via the cabinet, the EGC (PE), the GEC, the grounding rod $R_{r,c}$, planet Earth and the grounding rod $R_{r,s}$. The voltage between cabinet and planet Earth is determined by the total resistance in the grounding system. Since planet Earth is part of this resistance, it is very difficult to achieve very low values, therefore TT systems are typical limited to situations where fault currents remain below 100 to 220 A. This equals a nominal connection value of 16 to 32 A, which is insufficient for larger industrial installations. Another disadvantage of the TT system is the occurrence of overvoltage between neutral and ground in the main distribution panel during a lighting stroke.

2.1.2 IT grounding system

The IT system is shown in Figure 2.2. In this situation the neutral does not have a direct connection to planet Earth (or only via a high impedance) and all phases are floating. A single fault

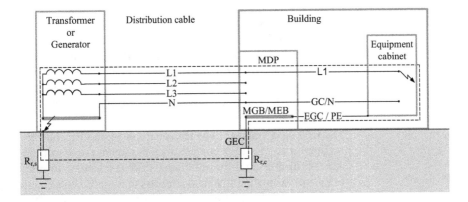

FIGURE 2.2 In an IT system all phases float. Such a solution is usually not chosen for the entire installation, but only for a critical section.

does neither lead to a dangerous situation nor to a shutdown of the system. IEC 60364-4-41 allows continuation of operation during a single fault; however, it also requires that provisions are taken to avoid harm to persons in case a second fault occurs. These provisions are relatively expensive. Since these are based on current monitoring, the allowed amount of ground leakage current is limited which is not viable for large installations. Therefore, IT systems typically are only used for those parts of an installation which require high availability. In addition, IT systems are used in medical treatment areas where already small ground leakage currents may be harmful for the patient.

2.1.3 TN grounding system

For large installations a TN system is preferred. Like in TT systems the star point of the mains transformer, or generator, is connected to planet Earth; however, the equipment cabinets are also connected to the system ground point. In fault situations, there is no current flow through planet Earth and the system can be dimensioned such that the voltage between the cabinet enclosure and soil remains small even for very large currents. The TN system knows three variants: TN-S (Figure 2.3a), TN-C (Figure 2.3b), and TN-C-S (Figure 2.3c).

2.1.3.1 TN-S system In the TN-S system, the PE/EGC is extended beyond the main distribution panel toward the power source and connected there to system ground: separate conductors are used for neutral and PE/ECG. During a lightning strike an overvoltage between these two conductors can be induced; however, this voltage will be smaller than in a comparable TT system. Within the building itself, there is no further difference between the TT- and TN-S system.

2.1.3.2 TN-C system TN-C system is slightly cheaper than TN-S because the PE/ EGC and neutral/grounded conductor are combined into a single conductor (PEN). At first glance also the EMC performance might seem favorable, because no voltages can be induced between neutral and ground. In Section 2.3, however, we will discuss some major drawbacks.

2.1.3.3 TN-C-S system The IEC 60364-5-54 and IEC 60439-1 require for copper PEN conductors a minimum cross-sectional area of 10 mm² and for aluminum PEN conductors 16 mm². For small loads the PEN conductor has to be split in a separate neutral and PE/ECG. Because it is difficult to maintain TN-C system integrity over the life span of an installation; it is highly recommended to split the PEN conductor in the MDP and not

(a)

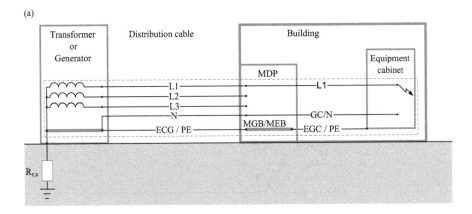

(b)

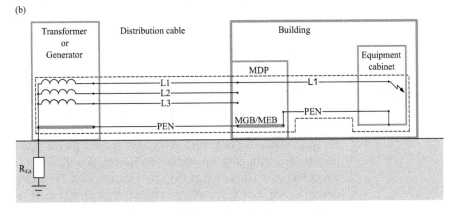

(c)

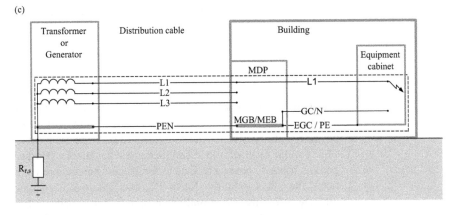

FIGURE 2.3 In TN systems a short circuit current (dashed line) returns via a conductor in the distribution cable to the power source: (a) TN-S; (b) TN-C; (c) TN-C-S.

Table 2.1   Grounding systems

	TT	IT	TN-S	TN-C	TN-C-S
Overvoltage in MDP during lighting strike	Y	Y	Y	N	N
Stray mains currents in PE/ECG	N	N	N	Y	N
Large short circuit currents in PE/ECG	N	N	N	Y	N
Large harmonic currents in PE/ECG	N	N	N	Y	N

to distribute it further to subpanels or loads. De facto this is a combination of the TN-C and TN-S system; hence, the naming TN-C-S is used. For healthcare installation in the United States such a system is mandated by the NPFA 70, because insulated ECG's are required.

2.1.4 **Comparison**   In Table 2.1, we summarize the positive and negative aspects of the discussed power distribution. We added the risk for large short circuit and large harmonic current flow over the PE/ECG. They not only have a negative impact on the EMC performance, but pose a potential fire hazard as well. In some countries, it is allowed that the cross-section of a PEN conductor is only half of the phase conductor, great care must be taken here. Overall, TN-C-S has the best performance, but for small or safety–critical installations, TT and IT, respectively, are acceptable.

2.2   Lighting protection

Lighting protection is an example where planet Earth is an essential part of the grounding system. In a thunder cloud positive and negative particles are separated from each other. The bottom side of the cloud becomes negatively charged and the top-side positively. Due to forces between charged particles the surface of the Earth becomes positively charged directly below the cloud and negatively charged further away from the cloud (see Figure 2.4). A large potential difference arises between the cloud and planet Earth until a lightning flash occurs. At that moment, a current will flow from the surface of planet Earth to the bottom of the cloud. This causes current flow in planet Earth itself. At the edge of the cloud current will

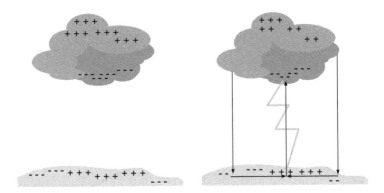

FIGURE 2.4 During a lightning strike, the soil is part of the current return path as indicated by the arrows.

flow from cloud to planet Earth, without an actual conductive path. This current is called displacement current, similar to the current in a capacitor during charging and discharging. In case cables have been buried in the ground, part of the lightning current may flow over them and cause overvoltages in the distribution network or cause problems in communication lines.

As Benjamin Franklin already demonstrated in 1752, with a kite and a key, a lightning stroke can be rerouted. In the modern version, lightning rods capture the stroke and give the large lightning current a safe path to planet Earth. When the current distributes over the Earth's surface, voltage drops will arise between various points. Therefore, the use of separate, unconnected, grounding rods for various parts of the grounding systems, as shown in Figure 2.5a, may lead to dangerous voltages between these parts (often referred to as *ground potential rise* or *transferred earth potentials*). The IEC 62305 prescribes the connection of lighting ground, safety ground, and galvanic tubing at least one point as shown Figure 2.5b. This point is called the *equipotential bonding bar.* In case galvanic bonding is not possible, the use of a *surge arrester* is mandatory. Actually, this means that in case overvoltages occur all groundings systems are connected to each other anyway. Since galvanic connections are more fail-safe (and cheaper), they are better preferred.

2.3 Bonding networks

Lightning rods have been successfully applied to prevent fire and damage to buildings for a long time. To protect sensitive electronics additional measures have to be taken.

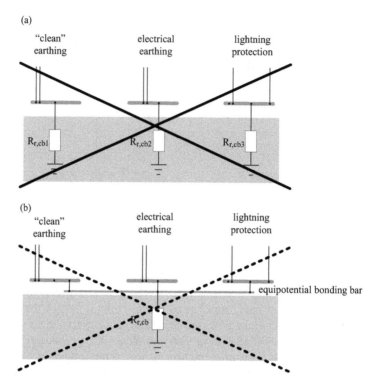

FIGURE 2.5 (a) Separate, unconnected, grounding electrodes are no longer allowed; (b) IEC 62305 requires equipotential bonding between various grounding systems. In case bonding is not possible, connection via surge protection is mandatory.

2.3.1 Common bonding

IEC 61000-5-2 advises a three-dimensional multiple bonding network, as shown in Figure 2.6. This network combines lightning protection, safety, functional and EMC grounding. It is often referred to as *MESH-CBN (Common Bonding Network)*, *MESH-BN*, or *CBN* derived from for *mesh common bonding/ bonded network*. It includes the use of the already available reinforcement of steel. In a case study for lightning protection of pharmaceutical plant, it was found by both measurement and simulation that 80% of the lightning current was carried by the roof structure and only 20% by the intended lightning conductors (Bargboer 2010).

2.3.2 PEN conductors

In Figure 2.7, a CBN is combined with a TN-C power distribution feeding two cabinets. The current delivered by phase L1 to cabinet 1 not only has a ground return path via its PEN conductor, but through other grounding connections as well. Though

(a)

Building

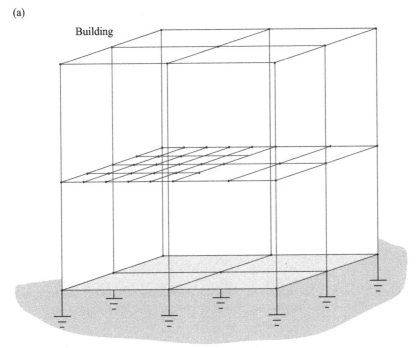

(b)

Building

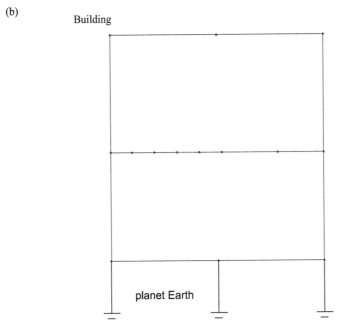

planet Earth

FIGURE 2.6   Mesh CBN; (a) 3D view; (b) 2D view.

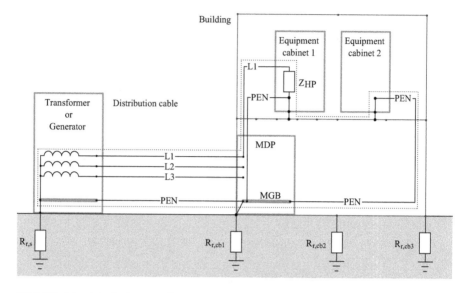

FIGURE 2.7 Mesh CBN with TN-C power distribution may lead to stray currents (dotted line). This design may be prohibited by national and local regulations.

most current will return through its PEN conductor, stray currents will flow through the grounding mesh and through the PEN conductor of the second cabinet (as indicated by the dotted line).

The proper solution is to employ a TN-C-S distribution where the PEN conductor is split into an EGC/PE and a grounded conductor/neutral in the mains distribution panel. The grounding system design shown in Figure 2.8 is not known to create safety conflicts with National Electric Code (NEC) and related Nationally Recognized Testing Laboratory (NRTL)-listed equipment installation requirements (IEEE 1100 1999) or international standards.

2.3.3 **Isolated bonding**

For new designs MESH-CBN should be applied, unless it connects equipment that is highly sensitive for minor variations in the magnetic field or if ground leakage currents have to remain very small for safety reasons. In telecommunications industry, some equipment manufacturers still require a single-point ground, often related to the use of DC power distribution. This is referred to as *isolated bonding network* or *IBN* (IEEE 1100 1999). IBN networks have either a star shape or form a local mesh (mesh-IBN) and normally are embedded in a CBN for the rest of the installation. Such a combination is called a *hybrid bonding network* (IEC 61000-5-2 1997). In either case,

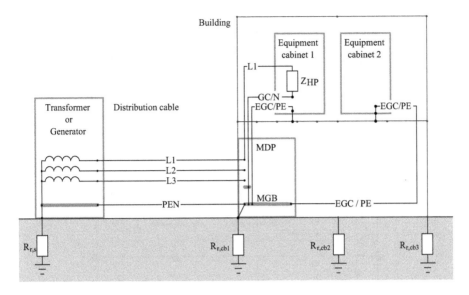

FIGURE 2.8 Mesh CBN with TN-C-S. This design is not known for causing conflict with national and local regulations.

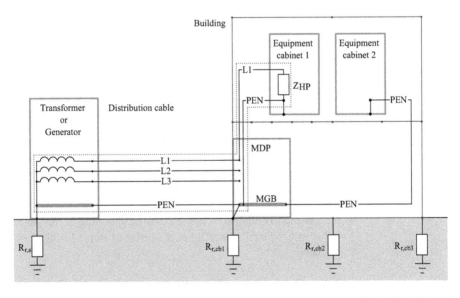

FIGURE 2.9 Combination of a CBN and an IBN where the mains grounding bar is the SPC.

the IBN is connected to the CBN at exactly one point. This point is called *single-point connection* or *SPC* (see Figure 2.9).

Also medical treatment areas typically employ a star-shaped grounding system, because already a very small, low frequency, leakage current through a patient may cause significant harm

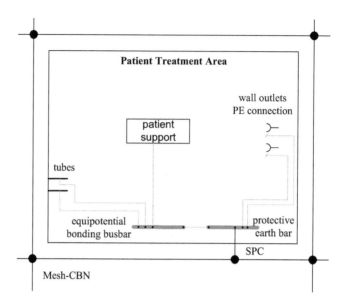

FIGURE 2.10 A star-shaped grounding systems is mandatory in medical treatment areas.

or even death. Such a system is implicitly described in IEC 60364-1 and may be enforced by National regulations. Outside the treatment area again the MESH-CBN is preferred (see Figure 2.10).

Another example can be found in the new fusion generator, ITER, a MESH-CBN is implemented for the whole building expect in the containment area where the plasma will be floating. This area is called a loop control zone and where feasible only a single-point grounding path is used. The rationale here is that the plasma is contained by magnetic fields. Already a very small disturbance in the intended magnetic field will cause the plasma to collapse. When discussing shielding of magnetic fields in Chapter 7, we will see a similar example for electron microscopy.

2.3.4 Resonance in bonding systems

The implementation of an IBN is not without risk. Sufficient isolation between the separately grounded parts (i.e., cabinets 1 and 2 and the CBN) have to be maintained over the full lifetime of the installation and under all circumstances. If the isolation is too small, parasitic capacitances will cause resonances excited by harmonic and lighting currents. In case of severe external disturbances, like a lightning strike, even a flashover may occur. Due to its vicinity to sensitive electronics, this may

FIGURE 2.11   Flashover between IBN (cabinet) and CBN (integrated in the floor).

cause more damage than the initial disturbance. In addition such a flashover may frighten people (Figure 2.11).

The demonstration in Figure 2.12 uses two small "cabinets" which both have their own ground connection. To simulate a disturbance source, we connect both wires to different poles of a function generator. The current through the wires is measured with a current probe and an oscilloscope. In the first experiment, the distance between both cabinets is large; the insulation proves to be sufficient, because no current is flowing. In the second experiment, both cabinets are only isolated with a thin piece of paper. Current starts flowing showing that we incidentally build a plate capacitor (C). Together with the self-inductance (L) of the loop formed by the ground wires a large current flows at the resonant frequency (f):

$$f = \frac{1}{2\pi\sqrt{LC}} \tag{2.1}$$

In the third measurement, we directly placed the cabinets on top of each other thus providing galvanic contact. As expected current flows in the common mode circuit; however, its amplitude is less than in the previous, resonant, measurement.

2.4   Ground as signal return

As discussed in the introduction of this chapter, ground is also used to refer to signal return paths. These return paths may or may not coincide with the grounding systems discussed previously.

(a)   (b)

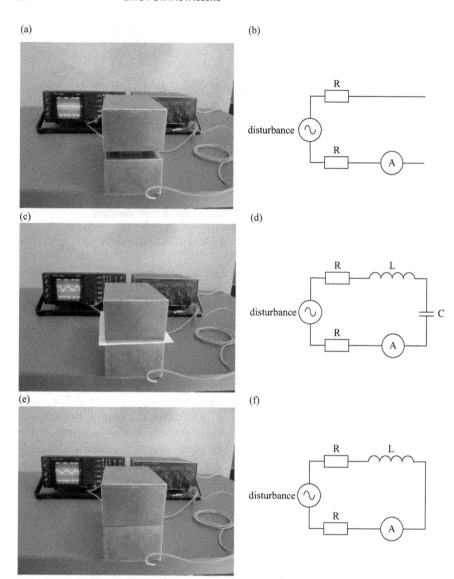

FIGURE 2.12 Resonance in grounding systems: (a + b) open loop due to the large distance between the cabinets; (c + d) resonant loop due to the capacitance between the cabinets; (e + f) short circuited loop because the cabinets are in contact.

2.4.1 **Signal return via CBN connection**

Figure 2.13a shows a close-up of the CBN installation from Figure 2.10. Cabinet 1 communicates with cabinet 2 via a single-wire interface. This implies that the transmitted signal returns via a common ground path. Schematically this is shown in Figure 2.13b.

(a)

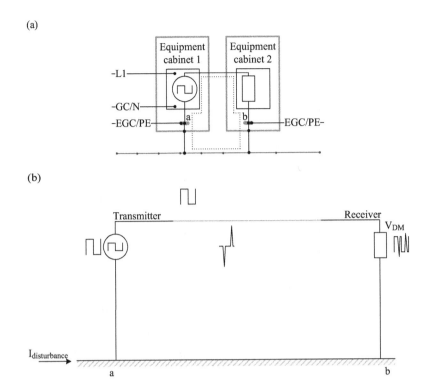

(b)

FIGURE 2.13 Ground as return conductor: (a) Single-wire interface between cabinet 1 and 2; (b) simplified representation. In both overviews, points *a* and *b* correspond to each other.

Currents through the grounding system will interfere with the communication signal. The voltage V_{DM} detected by the receiver in cabinet 2 is the sum of the originally transmitted voltage and the disturbance voltage. In Figure 2.14a, a test setup is shown schematically mimicking the common ground conductor (Helvoort 1994). A 10A current is sent through a copper tube. A voltage meter, V_1, is connected to this tube at points *a* and *b*. The measured voltage as function of frequency of the disturbance current is shown in Figure 2.14b. Only up to 70 Hz the measurement results are in line with Ohm's law:

$$V_1 = I \cdot R_{DC} = 2.6 \,[\text{mV}]\,\text{with}\, R_{DC} = 0.26 \,[\text{m}\Omega] \qquad (2.2)$$

Above 70 Hz the measured voltage V_1 increases with frequency. The voltage V_1 is often referred to as *ground potential rise* (or if the tube would be planet Earth itself *transferred earth potentials*). These terms have led to the widespread and persistent

(a)

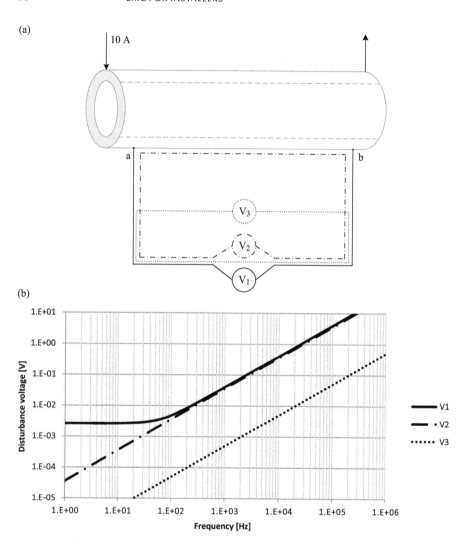

FIGURE 2.14 (a) Demonstration setup for measuring disturbance voltages caused by a current over a grounding conductor; (b) Voltage measurements, even without galvanic connection to the grounding conductor large voltages are measured at high frequencies.

misconception is that the increase in voltage is caused by the inductance of the ground wire. In nearly all situations, this increase is actually caused by mutual inductance: The current that is sent through the tube causes a magnetic field, this field induces a voltage which is measured as part of V_1, just like in an air coupled transformer. This can be demonstrated by connecting a volt meter V_2 to a loop which coincides with the loop formed by the wires connecting V_1 to the tube without

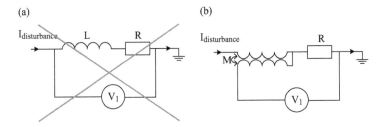

FIGURE 2.15 In the classical equivalent circuit (a) the current through the common impedance formed by a resistance and inductance causes an interference voltage. In the improved equivalent circuit (b) a transformer is used to model the high frequency behavior.

connecting it electrically to the tube. The voltage drop across the tube itself can therefore not be measured. Even without galvanic contact the measured voltage V_2 above 70 Hz nearly equals V_1, see Figure 2.15 for the appropriate network model.

In actual installations, the dimensions of loops are often much larger than the loops in this demonstration model (the largest loop measures 1 m × 0.26 m). Even at 50 Hz magnetic fields are often more important than potential differences between two distant points on a conductor.

2.4.2 Signal return via IBN connection

In Figure 2.16, we replaced the CBN of the cabinets in Figure 2.13 by a local isolation bonding network. Under the assumption that no disturbance currents will flow over the grounding conductors of both cabinets, this concept looks promising at first glance. However, we already have seen in the test setup of Figure 5.14 that the voltage V_2 induced in the galvanically unconnected loop is equivalent to V_1 at higher frequency. The mutual inductance contribution is more important than the resistive contribution, as becomes clear from the equivalent circuit in Figure 2.17. Thus, an important conclusion is that an IBN does not automatically perform better than a CBN even at low frequencies!

2.4.3 Single-point signal grounding

Further investigation of the equivalent circuit of a single-wire interface with IBN return shows that the disturbance can be lowered by lowering the mutual inductance. This can be achieved in two manners. First, the loop area of signal wire and return can be reduced. Second, the distance between the loop and the current carrying ground conductor can be increased. This is proven in the setup from Figure 5.14. The voltage, V_3, measured in the small loop, is 80 times lower than V_2, measured in the

(a)

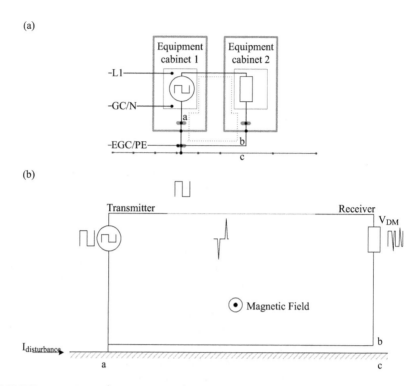

(b)

FIGURE 2.16 Ground as return conductor in an IBN setup: (a) Interface wire between cabinets 1 and 2; (b) simplified representation. In both overviews, points *a* and *b* correspond to each other.

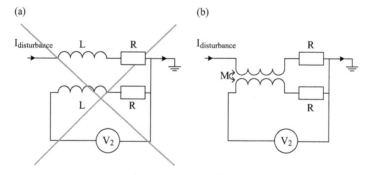

FIGURE 2.17 Equivalent circuit of a single wire interface with IBN return.

large loop. Figure 2.18 shows a possible implementation, where the grounding structure has been extended with a dedicated ground return for the signal. Such an implementation is called *single-point grounding*. The equivalent circuit in Figure 2.17 remains valid.

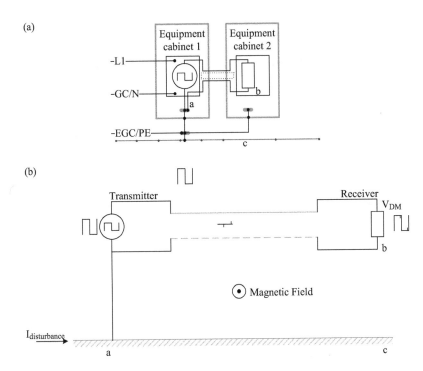

FIGURE 2.18 (a) Single-point ground of the communication circuit in a local IBN setup; (b) simplified representation. The EG of cabinet 2 is not shown in the latter.

2.4.3.1 Twisting In single-point grounding, mutual induction can be reduced further by twisting the signal and its return wire into a twisted pair as shown Figure 2.19a. Assuming perfect symmetry and a homogeneous magnetic field the voltage induced in each loop will be exactly opposite to the voltage in its neighboring loop and therefore will cancel each other (Figure 2.19b).

2.4.3.2 Resonance Parallel to the discussion and demonstration in Section 2.3.4, (parasitic) capacitance between points *b* and *c* may cause resonances as shown in Figure 2.20. Closer inspection shows that we actually introduced two loops. The small, closed, loop formed by the signal and return conductor is called the *Differential Mode* (DM) loop. The large, open, loop formed by the return and the current carrying ground conductor is called the *Common Mode* (CM) loop. The voltage which was present as disturbance in V_{DM} in Figure 2.16 is now induced in the CM loop and may drive a large current over the signal-return wire, which leads to a large disturbance in the received signal.

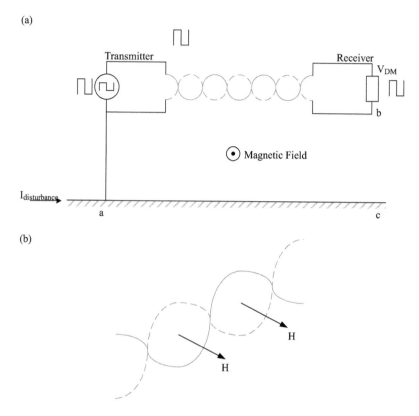

FIGURE 2.19 (a) Twisting reduces the magnetic coupling, because (b) induced voltages in neighboring twists cancel each other.

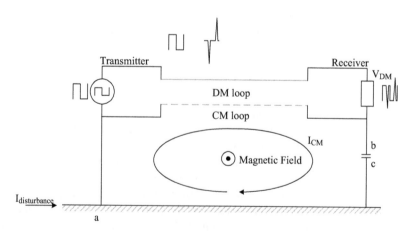

FIGURE 2.20 (Parasitic) capacitances in single-point grounding systems cause resonances and under severe circumstances lead to flashovers close to the sensitive receiver.

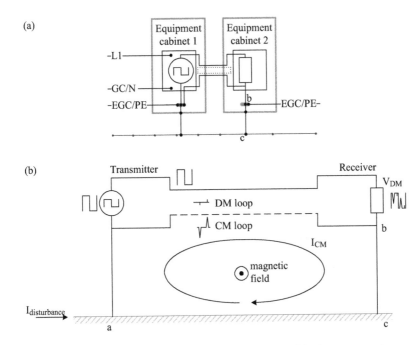

(a)

(b)

FIGURE 2.21   (a) Multi-point grounding in a CBN; (b) simplified representation.

2.4.4 Multi-point signal grounding

Both in Sections 2.1.3 and 2.2 overvoltage problems between ground conductors were solved by bonding them together. In Figure 2.21b, we have bonded points *a* and *b* together. This is called *multi-point grounding*. In essence, it is equivalent to the CBN approach and as such integral part of it as shown in Figure 2.21a.

2.4.4.1 DM current flow   Though multiple grounding paths exist in a multi-point grounding approach, most signal current will return via the nearby signal return conductor. Figure 2.22 shows a test setup and diagram, adapted from (Goedbloed 1990) to clarify this statement. A long parallel pair cable is on one end short circuited and on the other end connected to a function generator. With a function generation, a current is injected in conductors (2) and (3) via conductor (1). In comparison with Figure 2.21 conductor (2) is the signal return and conductor (3) is the PE/ECG, the conductors (2) and (3) correspond to the CM loop.

It is obvious that the resistance of conductor (3) is much lower than the resistance of conductor (2), yet measurements shows that hardly any current flows of conductor (3) and most returns

(a)

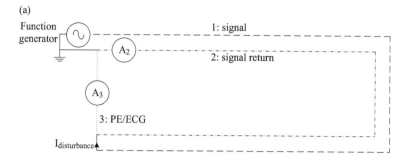

(b)

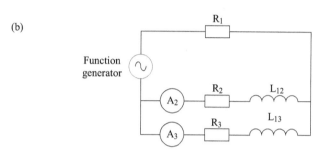

FIGURE 2.22 Current does not flow the path of least resistance, but of least impedance. The loop formed by conductors (1) and (2) is higher in resistance than the loop formed by conductors (1) and (3), but the self-inductance is lower.

to the source via conductor (2). The electrical loop formed by conductors (1) and (3) is much larger than the loop formed by the conductors (1) and (2). Therefore, the self-inductance (L_{13}) of the first loop is much larger than the self-inductance (L_{12}) of the second loop. At higher frequencies conductor (2) provides the path with lowest impedance.

*2.4.4.2 **Twisting*** Twisting in single-point grounding reduced the induced disturbance voltage. In multi-point grounded systems this is typically not the case, because the CM current is the dominant contributor. The direction of the magnetic field (*H*) caused by the CM current is opposite in neighboring twists; therefore, the induced voltages no longer cancel but add up as if the cable was not twisted at all. This is graphically clarified in Figure 2.23.

2.5 Reference installation

Based on the previous sections, a reference installation can be defined which should be the starting point for any new design

(a)

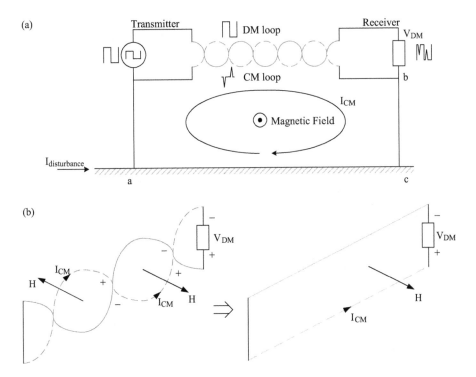

(b)

FIGURE 2.23 Twisting does not reduce disturbance voltages caused by CM currents.

or major renovation. Following the advice of IEC 61000-5-2 a mesh-CBN is implemented in combination with a TN-C-S power distribution, where the PEN conductor is split only in the MDP. After the MDP only ECG/PE and GC/N conductors are used. An implementation example is given in Figure 2.8. As feeder preferably a quad-core cable is used. For signal return paths the multi-point grounding concept is adopted. The overall scheme, which is not known to infringe any national or local regulation (IEEE 1100) is shown in Figure 2.24.

Bonding all grounding conductors together reduces the risk of resonances and flash-overs at the cost of the occurrence of a disturbance voltage in the DM circuit caused by currents in the CM circuit. The next three chapters discuss measures how to keep this disturbance voltage sufficiently low.

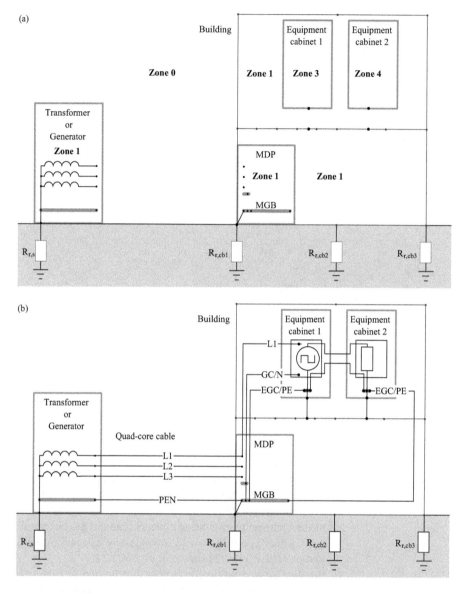

FIGURE 2.24 Reference installation with good EMC properties and not known to infringe any regulations: (a) zoning (see Chapter 1); (b) power distribution and grounding implementation.

CHAPTER THREE

Cabling, connectors, and assemblies

Many electromagnetic compatibility (EMC) noncompliances are due to the use of incorrect cabling or incorrect grounding of these cables. For instance, a failure in connecting the signal return wire in the reference installation of Figure 2.24, normally will not lead to immediate communication problems. During lighting storms or power fault conditions, the disturbances can be so severe that communication is hampered or components even damaged.

This chapter discusses unshielded and shielded cables, differential pairs, balancing, and EMC performance characterization.

3.1 Unshielded cables

In practice, most disturbances are caused by inductive coupling. A setup to show the vulnerability of a cable configuration to inductive coupling is shown in Figure 3.1. A current source, which can be switched 56 kHz, 570 kHz, and 4.5 MHz, injects approximately a 2 A current in two parallel bars above a ground plane (which represents a mesh-CBN). The resulting magnetic field is orthogonal to the center plane and zero exactly between both conductors. At this location a twisted pair cable with or without braided shield is mounted. At one end of this cable, its grounding configuration can be modified with differently wired connectors. On the other end, the disturbance voltage is measured with an oscilloscope. For broadband measurements instead of the current source and oscilloscope, a network analyzer has been used.

The cable configurations are labeled with letters, such that they correspond to the cable experiment described in (Ott 2009).

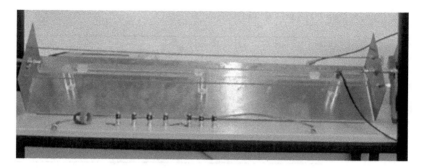

FIGURE 3.1   Setup for testing cabling configurations. Description and results are shown in Table 3.1.

Starting point for comparisons is a single wire above the ground plane (configuration A in Table 3.1), which is similar to Figure 2.13, thus mimicking a single return via a CBN. All voltages are normalized to a 1 A total current through the primary circuit.

Table 3.1 shows the measurements results of inductive coupling to unshielded cables. Configuration A is a single wire with the ground plane as signal return (see also Figure 2.13). The measured disturbance voltage increases linearly with frequency as predicted by the equivalent circuit in Figure 2.15, given that R is negligible: $|V| = 2\pi f M |I|$. Configuration D (see also Figure 2.21) shows that the voltage can be lowered by applying a parallel pair. For high frequencies, the voltage is lowered by roughly a factor 5. The single-point grounded configuration H (see also Figure 2.19) even seems more promising than the multipoint grounding of D. However, the broadband measurements in Figure 3.9 show that a resonance occurs near 40 MHz and the disturbance becomes much larger than in configuration A and D. This is another confirmation that resonance in grounding systems can lead to large voltages (see also Figure 2.20). Again, for longer cables, resonances occur at lower frequencies, e.g. the cable in the setup is 1 m long. If it had been 10 m long, the resonance would have been very close to our test frequency.

The difference between parallel pair and twisted pair wires has been discussed in Section 2.3.3 and 2.3.4. If the main disturbance is caused by CM currents, there is no difference in induced voltage between parallel pair and twisted wires.

3.2 Shielded cables

3.2.1 Coaxial cables

In coaxial cables a cable shield is used as return conductor as shown in Figure 3.2. Especially at higher frequencies, this is

Table 3.1 Test results of the cable configurations shown in Figure 3.1

ID	Configuration	Disturbance voltage (mV)		
		56 kHz	570 kHz	4.5 MHz
A	Signal return via ground plane.	19	290	1400
B	Coaxial cable. Shield grounded galvanic at one end and capacitive at the other end. Load connected to ground plane.	19	560	17
C	Coaxial cable. Shield grounded at both ends.	5.0	3.7	2.5
D	Signal return via wire grounded at both ends.	5.6	62	280
E	Signal return via wire grounded at both ends. Shield grounded galvanic at one end and capacitive at the other end. Load connected to ground plane.	5.6	100	12
Es	As E, capacitor shorted with pigtail.	3.4	18	12

(*Continued*)

Table 3.1 (*Continued*)   Test results of the cable configurations shown in Figure 3.1

ID	Configuration	Disturbance voltage (mV)			
		56 kHz	570 kHz	4.5 MHz	
Ef	As E, capacitor shorted with pigtail. Ferrite between load and ground plane.	1.5	4.6	2.8	
F	Signal return via wire grounded at both ends. Shield grounded at both ends. Load connected to ground plane.	3.3	4.6	2.6	
G	Coaxial cable, shield grounded galvanic at one end and capacitive at the other end. Load connected to shield.	0.09	13		
H	Signal return via wire grounded at one end.	0.018	0.95	39	
I	Signal return via wire grounded at both ends. Shield grounded galvanic at one end and capacitive at the other end.	0.012	0.26	0.51	
J	Signal return via wire grounded at one end. Shield grounded at both ends.	0.0099	0.099	0.52	

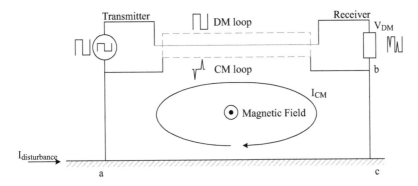

FIGURE 3.2   In coaxial cables the cable shield is used as signal return conductor.

an effective manner to further reduce the disturbance voltage. This can be shown with the setup in Figure 3.3. At higher frequencies the disturbance voltage drops exponentially (a physical explanation will follow in Section 3.4).

In Table 3.1, configuration C shows the test results of a coaxial cable in the setup of Figure 2.1. At 4.5 MHz, the disturbance voltage is reduced by almost a factor 200 when compared to a twisted pair (configuration D). Strictly speaking, the cable under test is not coaxial, because the inner wire meanders; however, in the given setup, its behavior is similar.

Coaxial cables can also be used in a single-point grounded system as shown in configuration G. Even though we added a capacitor between shield and ground, such that resonance occurs at 570 kHz, the disturbance voltage remains reasonably low, though it is 3.5 times higher than configuration C, it is more than a factor of 20 lower than the unshielded cable (configuration A). At higher frequencies, the behavior will be similar to configuration C, but this will depend on the quality of the capacitor. Care has to be taken that the signal and return are connected to the shield both at source and load. Configuration B is incorrectly connected as is clear from that the measurement results. At 570 kHz, the disturbance voltage is nearly thrice the voltage of the unshielded cable in configuration A.

3.2.2 **Bonding of shielded feeder cables**

Though from an EMC point of view, grounding both ends of a cable shield is preferred; sometimes, trade-offs with other aspects like cost are essential. Such an example can be found in power distribution networks. Instead of the quad-core cable between the mains transformer and the MDP as shown in the reference installation of Figure 2.24, single-core cables are preferred for high-power applications, because they are thinner and more flexible and therefore easier to handle.

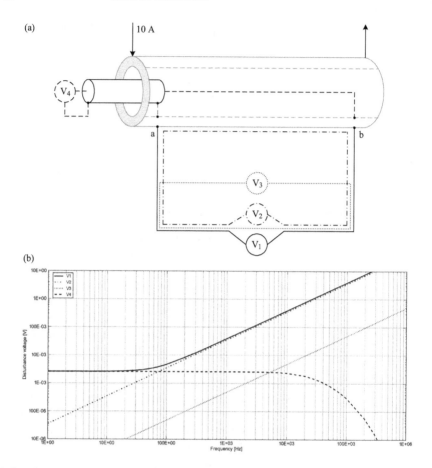

FIGURE 3.3 An extension of the measurement setup shown in Figure 2.14a shows that using a tubular shield is very effective in reducing the disturbance voltage: (a) setup where V4 represents a voltage measurement with the leads in the tube; (b) measurement result.

Single-core cables are provided with an integrated metallic shield which has to be grounded at one end at least to reduce the electric field outside the cable and to prevent (dangerous) capacitive current flow. Though normally not labeled as such, in essence single-core cables are coaxial cables. The large difference with coaxial cables used in signal transmission is that the resistivity of the shield is much larger than that of the inner conductor. Three different methods for bonding the ends to ground are in use: *both-ends bonding, single-point bonding,* and *cross-bonding* (ABB 2010).

3.2.2.1 Both–ends bonding   Figure 3.4 shows an arrangement in which both ends of the single core cables are bond to ground.

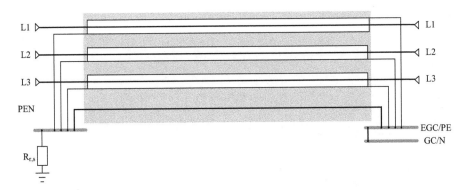

FIGURE 3.4   Both-ends bonding of single-core cables.

Under normal conditions most of the phase current will return through the shield. This can be proven by the setup in Figure 3.5, where 97% of the current returns via the shield and only 3% via the low resistive short between the shield ends. Actually this demonstration is a high-power version of the experiment shown in Figure 2.22: current flows via the path of least impedance which is not necessarily the least resistance.

Due to resistive dissipation in the shield additional heat will be generated in the cable, which will reduce the current carrying capacity of the cable. This reduction will be larger in a flat formation than in a trefoil, triangular, formation (see Figure 3.6). From an EMC view, on the one hand, the stray magnetic field will be reduced when both-ends bonding is applied; on the other end, stray mains currents can flow through other ground connections.

3.2.2.2 Single-point bonding   Figure 3.7 shows single-point bonded cables. The cable shields provide no path for currents and the current carrying capacity of the cable is maximal. However, with increasing length, an increasing voltage will build up between the shield and the ground at the open end. This effect limits the length of the cable stretch.

3.2.2.3 Cross-bonding   In a cross-bonding approach, the cables are sectionalized and cross-connected (Figure 3.8). Though both ends of the cable shield are grounded, no significant currents will flow over the electrically continuous metallic shield. Voltages will be induced between shield and ground which are maximal at the shield interconnections. The number of cable shield sections determines the maximum voltage. This method allows to

(a)

(b)   (c)

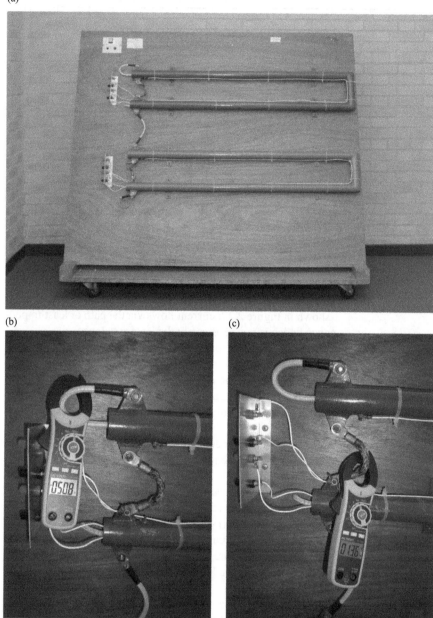

FIGURE 3.5 (a) The upper tube represents a single-core cable with both-ends bonding; (b) the current through the inner conductor is 50 A, and 48.7 A returns via the shield; (c) only 1.3 A returns via the short between the shield ends.

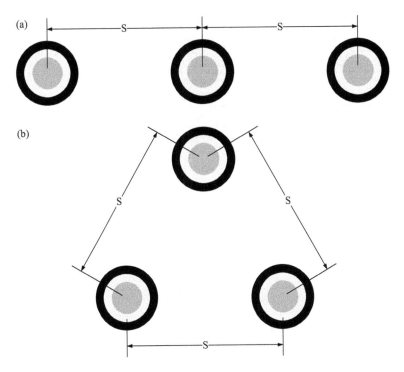

FIGURE 3.6 Single-core cable formations: (a) flat; (b) trefoil.

maximize cable current carrying capacity with longer routing lengths than with single-point bonding. The disadvantage with respect to both-ends bonding is that the stray magnetic field is not reduced and it requires additional hardware.

3.2.3 Shielded pair   In shielded pair cables the return conductor is conceptually separated from the shield. In reality, this depends very much on the exact connection. In case both the shield and the return wire are grounded at both ends (configuration F) the induced voltage is comparable to the coaxial cable (configuration C) over a very large frequency range as shown in Figure 3.9. At specific frequencies significant deviations, however, may occur.

For low frequencies and short cable lengths, configuration J provides an interesting alternative. Here the shield is grounded at both ends and the return conductor only at one end. Above 10 MHz both-ends grounding is preferred for the 1-m-long cable under test. For completeness, Table 3.1 lists also configurations E and I where the shield is grounded via a capacitor. Again resonances may cause large disturbances compared to galvanic bonded cables.

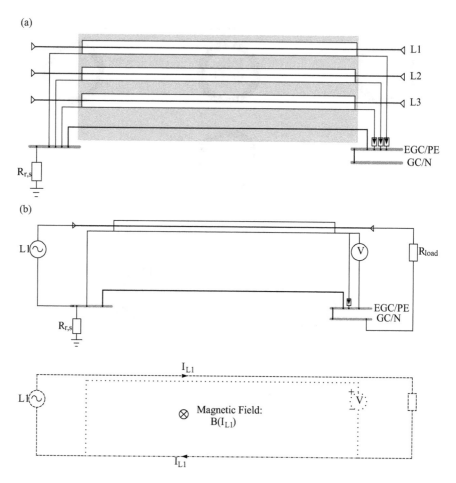

FIGURE 3.7 (a) Single-point bonding of single core cables; (b) single phase example: the current I_{L1} through the core (dashed loop) induces a voltage in the open loop (dotted) formed by the cable shield and the PEN conductor.

3.2.4 Pigtails

To maintain the EMC performance of a cable it is important that the shield is connected to ground over the full circumference. A proper example is shown in Figure 3.10a. An often used alternative is to unweave the strands of the shield and twist them together such that a short wire, called a *pigtail*, is formed. This pigtail is then connected to a ground pin or placed under a screw (see Figure 6.10b). Table 3.1 shows the negative effect of pigtails.

In configuration Es, the cable shield is connected to ground via a pigtail, while in F it is connected over the full circumference. The disturbance voltage in Es is significantly higher than in F. At high frequencies the performance of Es is not better than

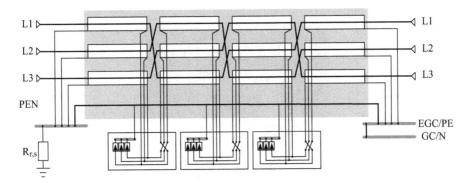

FIGURE 3.8 Cross-bonding of single core cables.

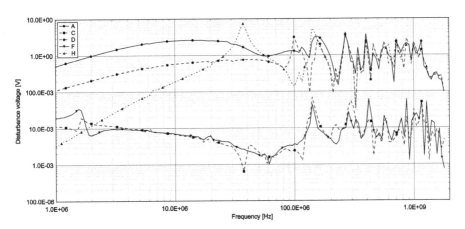

FIGURE 3.9 Broadband measurement on various cable configurations in the setup from Figure 3.1. The single point grounding of a parallel pair (H) is promising at low frequencies, but leads to a high resonance peak at a low frequency than grounding at both ends (D). Adding a shield to configuration D (F) is better and comparable to the behavior of a coaxial cable (C).

the unshielded cable (configuration E), because the capacitor provides a path with lower impedance than the pigtail. Nevertheless, for low frequencies using pigtails may be a cost-efficient solution (see Figure 3.11). Adding a ferrite in the connection between the pigtail and the return conductor ground can change the current paths such that the disturbance voltage is lowered as shown in Ef.

3.3 Differential pair

The previous sections considered only single-end signal transmission, which means that a signal is transmitted over a single

(a)

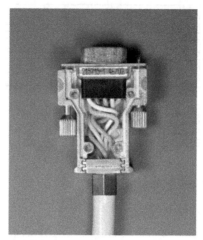

(b)

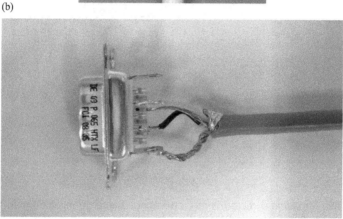

FIGURE 3.10   (a) cable connector which grounds the cable shield over its full circumference; (b) connector which uses a pigtail to connect the shield to ground.

wire and is referenced to ground. In this chapter, we consider differential signaling and balanced differential signaling.

3.3.1 Differential signaling

Configuration E employs an unshielded twisted pair in which one conductor is used for signal transmission and one is connected in parallel to ground. In differential systems both wires are used for signal transmission and referenced to ground. The original signal (V_t) is split in two and transmitted over both wires, but with opposite signs (V_1 and V_2) as shown in Figure 3.12. At the receiver the original signal is extracted by subtracting the received signals: $V_t = V_1 - V_2$. Actually, the original signal V_t is the voltage between both signal wires and is

FIGURE 3.11 Low-frequency application of pigtails: Bonding of power cables shields for lighting protection (as part of the CBN approach).

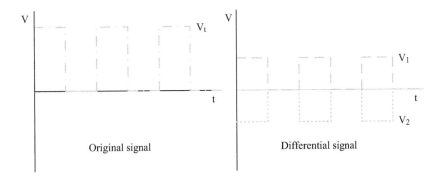

FIGURE 3.12 Differential signaling: $V_t = V_1 - V_2$.

called the differential signal. For this reason the term *differential mode* (DM), which was introduced in Section 2.3.3, is used for the loop formed by both wires.

Under the assumption that an external disturbance induces the same voltage, i.e., the *common mode* (CM) voltage, between ground and each of the signal wires, V_1 and V_2 can be measured separately. Even when both V_1 and V_2 are disturbed V_t can be restored as is shown in Figure 3.13.

Differential signaling is commonly used in fieldbus systems with parallel pairs. Active electronics is used to generate V_1 and V_2 from V_t on the transmitter site. At the receiver site both V_1 and V_2 are measured and electronically subtracted. Transceiver chips for a variety of protocols are available off-the-shelf.

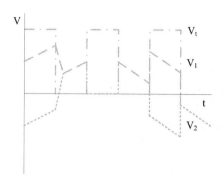

FIGURE 3.13   When both V_1 and V_2 are disturbed in the same manner, the transmitted signal V_t still can be detected reliably by subtraction.

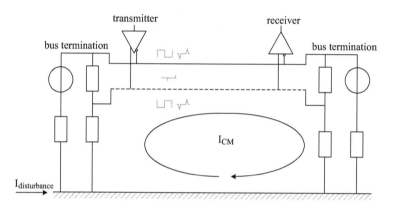

FIGURE 3.14   Industrial field busses use active electronics for differential signaling.

Figure 3.14 shows a schematic of an industrial fieldbus with differential signaling.

3.3.2 Balanced lines

The pull-up and pull-down resistors in Figure 3.14 connect the bus lines to ground. Therefore, a CM current can flow over both signal wires. In case there is unbalance between these wires, or their connections to ground, the CM current will distribute unevenly over these wires and will cause a disturbance voltage in the DM circuit. To prevent this, both circuits should be balanced, which in practice means making everything as symmetric as possible.

First, the active electronics can be replaced by passive transformers as shown in Figure 3.15. These types of transformers are called *baluns*, because they can convert *balanced* signals (i.e. V_1 and V_2) into *unbalanced* signals (V_t) and vice versa.

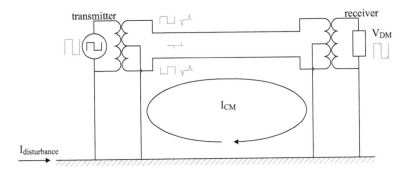

FIGURE 3.15   Differential signaling with baluns.

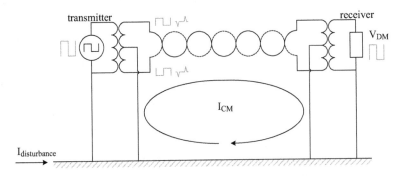

FIGURE 3.16   Twisting improves symmetry of the cable with respect to its environment.

Second, the signal wires can be twisted (see Figure 3.16). This improves the symmetry of the cable with respect to its environment. Furthermore, twisting reduces direct induction as already shown in Figure 2.19.

3.3.2.1 UTP, STP, FTP, and SFTP   Balanced twisted pairs are best known from twisted-pair Ethernet networks (IEEE 802.3). Typically, the balun's are integrated in the networking equipment, somewhat hiding the fact that twisted-pair Ethernet networks are intrinsically multipoint grounded systems, unless specific measures like isolation transformers are added.

For Ethernet networks a variety of cable types have been developed:

- Unshielded twisted pair (UTP)
- Shielded (braided) twisted pair (STP)
- Foil twisted pair (FTP)
- Shielded (braided) FTP (SFTP)

Over the past few decades, there has been a continuous discussion which cable type is best. This indicates there is no straightforward answer to this question. The reason for this is relatively simple: if an ideal UTP network could be realized in practice it would be disturbance free and any type of shielded cable would add unnecessary cost. Unfortunately the ideal network cannot be realized. First, components like baluns, connectors, and cables have tolerances. Second, during installation it is impossible to cut cable length at an exact integer and even number of twists. Third, during mounting it is necessary that part of the wiring is untwisted, while also the RJ-45 itself is not symmetric and compensation measures are necessary to restore balance. The symmetry is even further disturbed when cables are being pulled (too hard) instead of being laid. The unbalance in twisted-pair networks is the most important reason for disturbances. This unbalance is called *transverse conversion loss* (TCL) (IEC 11801 2009).

If the TCL is too large the induced disturbance can be reduced by adding a shield (see Section 3.2.3). For high data rate connections, which means high frequencies, great care should be taken when connecting the cable shield to ground (see Section 3.2.4). Under laboratory conditions it has been shown that improper connection can lead to higher radiation of an FTP cable compared to a UTP cable. From history it seems that at the introduction of higher speeds network equipment shielded cables are preferred, while when this new technology has matured unshielded cables can be used.

One of the arguments regularly used against using shielded twisted-pair cables is that the shield causes additional damping of the transmitted signal. Theoretically, this is correct; however, in practical installations, the same argument can be used against unshielded cabling, because cabling is typically routed in metal conduits in which case damping also occurs but in an undefined manner.

3.4 EMC characterization

A variety of properties are used to characterize the EMC performance of cables and cable assemblies, such as transfer impedance, shielding effectiveness, screening attenuation, and coupling attenuation, which will be discussed in the next paragraphs. In the past also optical braid coverage was used, but has proven to be not useful.

3.4.1 Transfer impedance

The *transfer impedance* (Schelkunoff 1934), sometimes also called *surface transfer impedance*, is the most robust method to quantify EMC performance. Its value can be determined by sending a CM current over the cable assembly and measure the induced differential voltage. Two different methods are in common use:

- Triaxial (Vance 1974)
- Wire injection (Eicher 1988)

The triaxial method is shown in Figure 3.17. The CM current is injected over the cable under test and returns via an outer tube. For coaxial cables the complete structure is triaxial, hence, its name. The triaxial setup is intrinsically shielded which is beneficial for both its emission and immunity. The wire injection method replaces the tube by a single wire as shown in Figure 3.18. To prevent unwanted emission typically the disturbance current is connected to the DM circuit and the voltage induced in the CM circuit formed by the shield and injection wire is measured. Care must be taken that no external

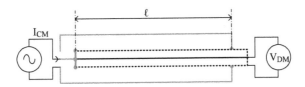

FIGURE 3.17 Schematic representation of the triaxial transfer impedance measurement method.

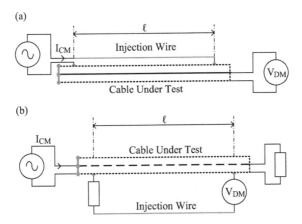

FIGURE 3.18 Schematic representation of the wire injection transfer impedance measurement method: (a) Injection wire on the outside; (b) injection wire used as detector.

voltages are induced. In addition, one has to take into account that the measured transfer impedance no longer may be an intrinsic property of the shield, but may depend on the position of the inner and outer wires, in case the cable is not coaxial (Helvoort 1995).

For connectors and other local shielding discontinuities, the transfer impedance (Z_t) is the ratio between the DM voltage and the CM current:

$$Z_t = \frac{V_{DM}}{I_{CM}} \ (\Omega) \tag{3.1}$$

For cables, the transfer impedance typically is specified per unit length (Z_T) and can be used in transmission line analysis (Degauque 1993):

$$Z_T = \frac{V_{DM}}{l \cdot I_{CM}} \ \left(\Omega/m\right) \tag{3.2}$$

The definitions of Z_t and Z_T is are conform to IEC TS 62153-4-1 (2014). The total transfer impedance of complete cable assembly l is given by

$$Z_{t,assembly} = Z_{t,connector1} + Z_{t,connector2} + l \cdot Z_{T,cable} \tag{3.3}$$

In Figure 3.20, the transfer impedance of a variety of cables, depicted in Figure 3.19, is shown.

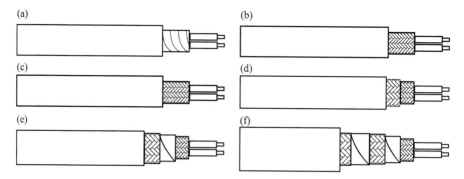

FIGURE 3.19 Cables with different types of shields. Cable types: (a) aluminized mylar; (b) single stranded braid; (c) single optimized braid; (d) double optimized braid; (e) super-screen; (f) double superscreen.

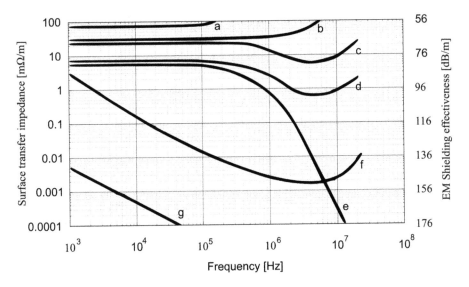

FIGURE 3.20 Transfer impedance measurement results for the cable types in Figure 3.19.

3.4.2 Transfer admittance and effective transfer impedance

Next to the transfer impedance, there is also a *transfer* or *coupling admittance* (Y_C), which specifies the amount of DM current caused by a CM voltage:

$$Y_C = j\omega C_T = \frac{I_{DM}}{l \cdot V_{CM}} \quad (1/\Omega m) \tag{3.4}$$

where C_T is the *through capacitance* per meter.

The measurement setup is schematically shown in Figure 3.21. In contrast to the transfer impedance the transfer admittance always depends on the shield, the geometry, and the dimension of the internal and external circuits: C_T is formed by the inner conductor (solid line), the outer most conductor (gray line), and openings in the cable shield (dashed black line).

In realistic implementations, the impedance of the CM circuit is not infinite and the impedance of the differential

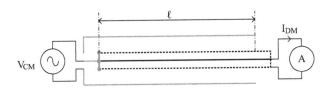

FIGURE 3.21 Schematic representation of triaxial transfer admittance measurement.

circuit will be larger than 0. In that situation, we can define the *capacitive coupling impedance Z_F (Ω/m)* (Vance 1975):

$$Y_C = j\omega C_T = \frac{I_{DM}}{l \cdot V_{CM}} = \frac{\dfrac{V_{DM}}{Z_{DM}}}{l \cdot Z_{CM} I_{CM}} = \frac{V_{DM}}{l \cdot Z_{DM} Z_{CM} I_{CM}} \Rightarrow$$

$$\frac{V_{DM}}{l \cdot I_{CM}} = j\omega C_T Z_{DM} Z_{CM} \ \left(\Omega/\text{m}\right) \Rightarrow \tag{3.5}$$

$$Z_F \triangleq j\omega C_T Z_{DM} Z_{CM} \ \left(\Omega/\text{m}\right)$$

In Figure 3.22, the equivalent circuit of the triaxial setup is shown with the addition of terminating impedances. In this case, the *effective transfer impedance* (Z_{te} or Z_{TE}) can be defined. Its definition, however, depends on whether the disturbance voltage is measured at the near (n) or at the far end (f):

$$Z_{te,f} = Z_f - Z_t \tag{3.6}$$

$$Z_{te,n} = Z_f + Z_t \tag{3.7}$$

For good quality shields, which are properly grounded, the capacitive coupling impedance is negligible (Benson 1992; Broyde 1993). The transfer impedance and admittance are reciprocal, which means that they do not only describe the coupling from an external circuit to disturbance voltages and currents in a cable, but also vice versa describe the disturbance a cable causes to the outside world.

3.4.2.1 Cross talk   *Cross talk* is at the border between EMC and *signal integrity* (SI). It means any interference from one communication channel to another. Figure 6.22 shows two signal wires above a common ground plane. The coupling between the

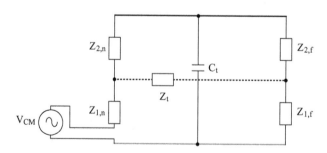

FIGURE 3.22 Equivalent circuit of the triaxial measurement setup.

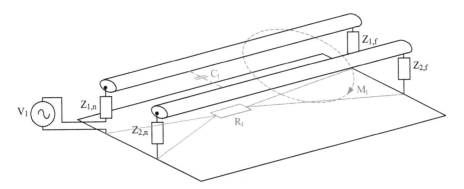

FIGURE 3.23   Coupling paths (in gray) between two signal wires above a ground plane.

wires is caused by the common resistance (R_t), the capacitive coupling (C_t), and the inductive coupling (M_t). Cross talk at the near end (*NEXT*) is typically larger than cross talk measured at the far end (*FEXT*). The equivalent circuit of Figure 3.23 is identical to Figure 3.22, and as deduced above, the resistive and inductive coupling reduce the capacitive coupling at the far end and add up at the near end. Further, for long cables, the signal will be attenuated more toward the far end, thus reducing the far-end cross talk.

3.4.3 Transmission lines

Electrical signals travel at high speed. When measurements on the source and load are made simultaneous, a time delay ΔT may be observed as shown in Figure 3.24. For sinusoidal signals this delay is ignored as long as it is less than 10% of the repetition time T. A cable or a wire above a ground plane with length l is called *electrically long* when the delay cannot be ignored. It is usually expressed in relation to the *wavelength* λ:

$$\Delta T = \frac{l}{v} = \frac{l}{\lambda}T \Leftrightarrow \frac{l}{\lambda} = \frac{\Delta T}{T} > 10\% \Rightarrow l > \frac{\lambda}{10} \qquad (3.8)$$

In this expression, v is the speed of the signal in a cable. For a wire or shield above a ground plane, it equals the *speed of light*, $c = 3 \cdot 10^8$ m/s. By definition, the signal frequency determines if the cable is electrically long:

$$f > \frac{v}{10l} \qquad (3.9)$$

A 1-m-long wire above a ground plane thus is electrically long above 70 MHz. A *transmission line* is a wire or cable construction which has been intentionally, or unintentionally, designed

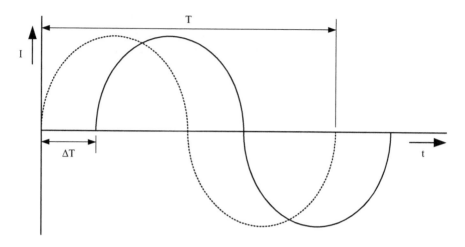

FIGURE 3.24 Time delay between source (dotted line) and load (solid line) for a wire above a ground plane.

to support signal transmission over electrically long conductors. The distance between the conductors in a transmission line is smaller than a tenth of the wavelength ($\lambda/10$).

*3.4.3.1 **Distributed voltages and currents*** Figure 3.25 shows an electrically long wire with the single sine pulse of Figure 3.24 propagating over it. As can be observed, the measured current (or voltage) in the *transmission line* does not only depend on the time, but also on the location at the wire. Both voltage and current move as a wave through the transmission line.

This phenomenon is known since the introduction of the telegraph and has been analytically described with a pair of

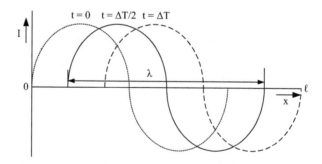

FIGURE 3.25 Current as measured on point *x* along an electrically long wire above a ground plane at three different moments in time.

differential equations known as the *telegrapher's equations* (Heaviside 1892). The telegrapher's equations can be modeled as an electrical circuit consisting of a ladder network with identical components in every section, such that every segment Δx is shorter than $\lambda/10$ (see Figure 3.26).

The components R', L', G', and C' are the distributed resistance, inductance, conductance, and capacitance, respectively, each representing an infinitesimal small length and are expressed per unit length. The equivalent diagram of the transmission line shows that the ratio of the voltage over each section and the current through each section equals

$$Z_0 = \sqrt{\frac{R + j\omega L}{G + j\omega C}} \tag{3.10}$$

where Z_0 is called the *characteristic impedance*. In the lossless case, Z_0 reduces to

$$Z_0 = \sqrt{\frac{L}{C}} \tag{3.11}$$

In measurement systems often 50 Ω coaxial cables are used, this means that these cables have a characteristic impedance $Z_0 = 50\ \Omega$.

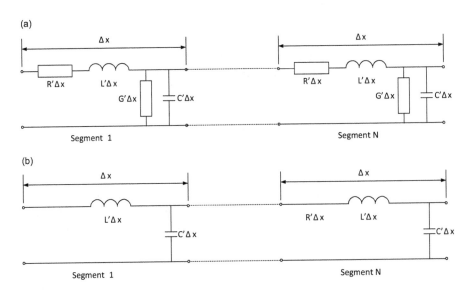

FIGURE 3.26 Equivalent diagram of (a) a lossy transmission line and (b) a lossless transmission line.

3.4.3.2 *Reflection, transmission, and cable resonance*   The solutions of the telegrapher's equations show that the voltage and current waves can be reflected at the end of the cable. The exact behavior depends on the impedance of the termination:

- Open
  - The voltage wave is reflected with positive sign.
  - The current wave is reflected with negative sign.
- Short
  - The voltage wave is reflected with negative sign.
  - The current wave is reflected with positive sign.
- Characteristic impedance
  - No reflection

In the *CBN* grounding approach typically shields are grounded at both ends. When the disturbance takes the shape of continuous wave, then reflections will add up to each other. If the cable length is an even times half the wave length (i.e., $l = n\lambda/2$) the voltages and waves amplify each other and *resonances* occur.

In general terms, the reflection can be described with the reflection coefficient ρ (Ramo 1984):

$$\rho = \frac{Z_L - Z_0}{Z_L + Z_0} \tag{3.12}$$

The reflected voltage wave V_r then is given by ρV_i, where V_i is the incident wave. The reflected current wave I_r is given by $-\rho I_i$. The total voltage at the load V_L and the total current I_L through the load are then given by.

$$V_L = V_i + V_r = (1+\rho)V_i$$
$$I_L = I_i - I_r = (1-\rho)I_i \tag{3.13}$$

Instead of a connection between a transmission line and a discrete impedance also, two transmission lines can be connected to each other. In this case it is useful to introduce the transmission coefficient τ:

$$\tau = 1 + \rho = \frac{2Z_L}{Z_0 + Z_L} \tag{3.14}$$

The incident voltage and current waves V_i, I_i from one cable then lead to voltage and current waves V_t, I_t in the second cable:

$$V_t = \tau V_i$$

$$I_t = (2 - \tau)I_i$$

(3.15)

3.4.3.3 Power measurements and S-parameters

The reflection and transmission coefficients can also be expressed in terms of instantaneous incident W_i, reflected W_r and transmitted power W_t:

$$\rho^2 = \frac{W_r}{W_i}$$

$$\tau(2 - \tau) = 1 - \rho^2 = \frac{W_t}{W_i}$$

(3.16)

With a *network analyzer* the incident and reflected power can be measured with a *directional coupler* as shown in Figure 3.27. This is sufficient to characterize a one-port device (such as a cable terminator). If the device under test (DUT) has two ports

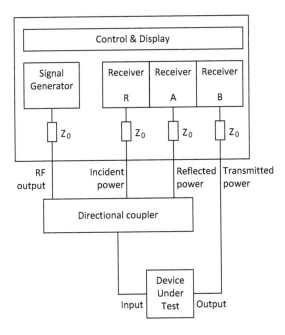

FIGURE 3.27 Network analyzer with directional coupler to determine reflection and transmission coefficients of DUT. Typically, Z_0 is 50 Ω.

the transmitted power can be measured directly if the network analyzer has sufficient receivers. The characteristic impedance used to connect the DUT to the analyzer must match the termination impedance of the signal generator and receivers. Mostly $50\,\Omega$ is used in EMC laboratories. This means that both the input and output of the DUT are loaded with $50\,\Omega$ during the measurement.

A more convenient way to characterize the DUT is to make use of an S-parameter test set. S-parameter test sets are available as add-ons to network analyzers (see Figure 3.28), or are fully integrated in a single unit. *S-parameters* or *scattering parameters* have been introduced by Kurokawa (1965). In a two-port system there are four S-parameters (Anderson 1967):

- s_{11}: the reflection coefficient ρ_{input} of the input port
- s_{22}: the reflection coefficient ρ_{output} of the output port

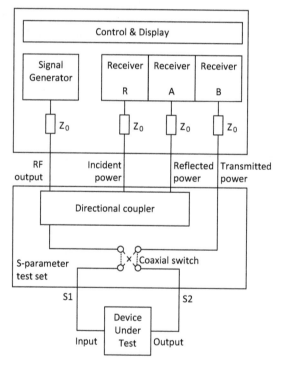

FIGURE 3.28 Network analyzer with S-parameter test set to determine reflection and transmission coefficients of DUT. The coaxial switch can swap the connections between the network analyzer and the ports S1 and S2. Other implementations, in particular for integrated units exist.

- s_{12}: gain G_{12} between input and output ($G_{12} = 20\log(s_{12})$ [dB])

- s_{21}: gain G_{21} between output and input ($G_{21} = 20\log(s_{21})$ [dB])

   In most EMC measurements, e.g., on cables and filters $s_{12} = s_{21}$. Both s_{12} and s_{21} can be negative, which actually means that the DUT provides attenuation.

3.4.3.4 Coupled transmission lines In Figure 3.2, a coaxial cable above a ground plane was shown. When the cable is electrically long the transfer impedance and admittance concepts remain very useful and describe the coupling of the transmission line formed by the cable shield above the ground plane (CM loop) and the transmission line formed by the cable itself (DM loop). The cable terminations and ground connections can be considered as four ports connected to the two *coupled transmission lines* (Goedbloed 1990) as depicted in Figure 3.29. An electrically short segment is shown in Figure 3.30. In the frequency domain, the transmission line equations of the four-port network are as follows:

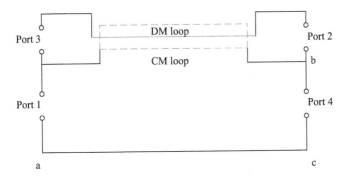

FIGURE 3.29 Coaxial cable above a ground plane depicted as four-port network.

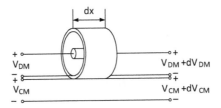

FIGURE 3.30 Electrically short segment of the coupled CM and DM loops from Figure 3.2.

$$-\frac{dV_{DM}}{dx} = Z'_{11}I_{DM} + Z'_{12}I_{CM} \tag{3.17}$$

$$-\frac{dV_{CM}}{dx} = Z'_{21}I_{DM} + Z'_{22}I_{CM} \tag{3.18}$$

$$-\frac{dI_{DM}}{dx} = Y'_{11}V_{DM} + Y'_{12}V_{CM} \tag{3.19}$$

$$-\frac{dI_{CM}}{dx} = Y'_{21}V_{DM} + Y'_{22}V_{CM} \tag{3.20}$$

Z'_{11} and Z'_{22} are the characteristic impedances of the uncoupled transmission lines and $Y'_{11}\ Y'_{22}$ are the characteristic admittances (i.e., the mathematical inverse of the characteristic impedance). The coupling between both transmission lines is described by Z'_{12}, $Z'_{21}Y'_{12}$, and Y'_{12}. Linear passive networks are reciprocal (Butterweck 1979); thus, $Z'_{12} = Z'_{21}$ and $Y'_{12} = Y'_{21}$. Per definition the transfer impedance and transfer admittance per unit length are as follows (Degauque 1993):

$$Z_T = -Z'_{21} = -Z'_{12} = \frac{1}{I_{CM}} \frac{dV_{DM}}{dx}\bigg|_{I_{DM}=0} \tag{3.21}$$

$$Y_T = Y'_{21} = Y'_{12} = -\frac{1}{V_{CM}} \frac{dI_{DM}}{dx}\bigg|_{V_{DM}=0} \tag{3.22}$$

Figure 3.31 shows the electrical equivalent schematic with

$$Z_T = R'_T + j\omega M'_T$$
$$Y_T = j\omega C'_T \tag{3.23}$$

For an electrically short cable, we can easily integrate Eqs. 3.17–3.20 over the length of the cable:

$$V_{DM} = \int_0^l \frac{dV_{DM}}{dx}dx = \int_0^l (Z'_{11}I_{DM} + Z'_{12}I_{CM})dx$$

$$= lZ'_{11}I_{DM} + lZ'_{12}I_{CM} = Z_{11}I_{DM} + Z_{12}I_{CM} \tag{3.24}$$

$$V_{CM} = \int_0^l \frac{dV_{CM}}{dx}dx = \int_0^l (Z'_{21}I_{DM} + Z'_{22}I_{CM})dx$$

$$= lZ'_{21}I_{DM} + lZ'_{22}I_{CM} = Z_{21}I_{DM} + Z_{22}I_{CM} \tag{3.25}$$

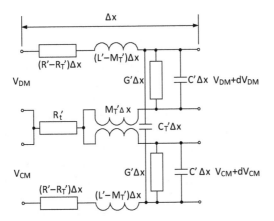

FIGURE 3.31   Circuit equivalent diagram for an electrically small cable segment.

$$I_{DM} = \int_0^l \frac{dI_{DM}}{dx} dx = \int_0^l \left(Y'_{11} V_{DM} + Y'_{12} V_{CM} \right) dx$$

$$= l Y'_{11} V_{DM} + l Y'_{12} V_{CM} = Y_{11} V_{DM} + Y_{12} V_{CM} \qquad (3.26)$$

$$I_{CM} = \int_0^l \frac{dI_{CM}}{dx} dx = \int_0^l \left(Y'_{21} V_{DM} + Y'_{22} V_{CM} \right) dx$$

$$= l Y'_{21} V_{DM} + l Y'_{22} V_{CM} = Y_{21} V_{DM} + Y_{22} V_{CM} \qquad (3.27)$$

The overall transfer impedance and admittance are then given by

$$Z_t = l Z_t = \frac{V_{DM}}{I_{CM}} \bigg|_{I_{DM} = 0} = Z_{12} = Z_{21} \qquad (3.28)$$

$$Y_t = l Y_t = \frac{I_{DM}}{V_{CM}} \bigg|_{V_{DM} = 0} = Y_{12} = Y_{21} \qquad (3.29)$$

The parameters Z_{12}, Z_{21}, Y_{12}, and Y_{21} can be conveniently determined with a two-port *S-parameter* set connected to ports 1 and 2 while short circuiting, respectively, leaving open ports 3 and 4.

$$Z_t = Z_0 \frac{2 s_{21}}{\left(1 - s_{11}\right)\left(1 - s_{22}\right) - s_{12} s_{21}} \qquad (3.30)$$

$$Y_t = \frac{1}{Z_0} \frac{-2s_{21}}{(1+s_{11})(1+s_{22})-s_{12}s_{21}} \qquad (3.31)$$

Theoretically, s_{21} and s_{12} are identical, but may show some variation in an actual measurement due to equipment tolerances. The term Z_0 in this equation stands for the characteristic impedance of the S-parameter set and its cabling.

To improve the high-frequency response, ports 3 and 4 can be terminated by the characteristic impedance of the respective transmission lines. In this case the effective transfer impedance is measured. For well-shielded cables, the admittance part can be ignored and the effective transfer impedance equals the transfer impedance as discussed in Section 6.4.2. A measurement setup is shown in Figure 3.32. Ports 1 and 2 are purposely on the opposite ends of the cable, because this gives the better frequency response (Degauque 1993).

3.4.3.5 Double-shielded cables   Figure 3.20 shows that double-shielded cables perform better than single-shielded cables. Typically, these shields are either short circuited (recommended) or left open. In both cases, reflections will occur in the transmission line formed by both shields, which may decrease the transfer impedance below the level of a single shield (Campione 2016).

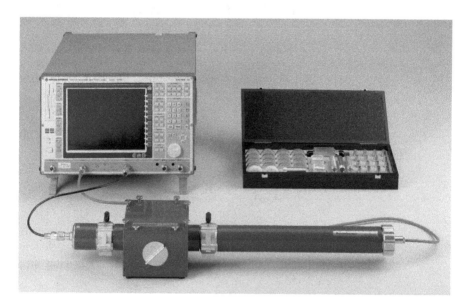

FIGURE 3.32 Triaxial test setup connected to S-parameter analyzer (reprinted with permission from Bedea).

3.4.4 Shielding effectiveness

Transfer impedance measurements become cumbersome at high frequencies due to multiple reflections caused by imperfections in the termination (see Section 3.2.3). For these frequencies the cable can be installed such that it acts as a wire *antenna* for an electromagnetic wave. The detected voltage by a cable without the shield will be higher than the measured voltage in the cable with the shield. The ratio between those two voltages is called *shielding effectiveness (SE)* (see Figure 3.33):

$$SE = 20\log_{10}\left(\frac{V_{received\ without\ shield}}{V_{received\ with\ shield}}\right) (dB) \qquad (3.32)$$

If a *mode-stirred chamber*, also known as *reverberation chamber*, is used to measure the shielding effectiveness, it is possible to correlate shielding effectiveness and transfer impedance.

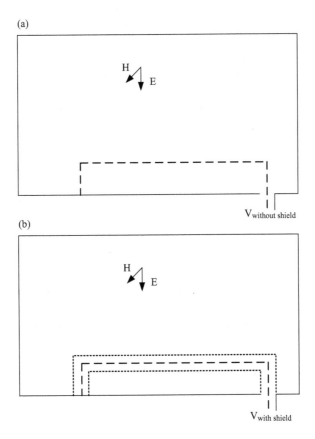

FIGURE 3.33 Shielding effectiveness is the ratio of (a) the measurement with shield to (b) the measurement without shield.

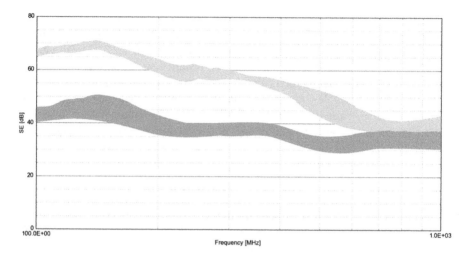

FIGURE 3.34 Example of shielding effectiveness of balanced cables: dark gray cat 5 UTP; light gray cat 5 FTP (both AMP systems cable).

A mode-stirred chamber is a resonant room in which a metallic stirrer changes the resonant modes in a controlled manner. The shielding effectiveness is averaged over all possible settings. An example is given in Figure 3.34. The relation between shielding effectiveness and transfer impedance has first been established by Martin (1982):

$$SE \text{ [dB]} = 36 - 20\log_{10}\left(Z_t\left[\Omega\right]\right) \tag{3.33}$$

This formula is applied in Figure 3.20. The given definition of shielding effectiveness can also be applied to balanced cables even if they are unshielded. In this case the balanced versus unbalanced situation is compared.

3.4.5 **Screening, unbalance, and coupling attenuation**

The definition of *screening attenuation* (a_s) is comparable to the shielding effectiveness and can also be determined in a *reverberation chamber* (IEC 61726 2015). Instead of using an unbalanced unshielded cable as reference, which requires extensive sample preparation, the total amount of power (P_{REF}) in the reverberation room is measured by a reference antenna. This then is compared to the amount of power (P_{DUT}) measured inside the shielded cable:

$$a_s = -10\log_{10}\left(\frac{P_{DUT}}{P_{REF}}\right) \tag{3.34}$$

DUT stands for *device under test*, which is used because this method can be applied to a variety of components.

If measurements are made on a balanced cable the attenuation is related to both the shield and to the unbalance. The ratio between P_{REF} and P_{DUT} then is called the *coupling attenuation* (a_c) which is the (logarithmic) sum of the screening attenuation (a_s) and the *unbalance attenuation* (a_u).

An alternative for a measurement in a reverberation room is the *absorbing clamp method* (IEC 62153-4-5 2013). In this case, the cable under test acts as transmitter. The configuration for measuring the coupling attenuation is shown in Figure 3.35a. The screening attenuation (a_s) can be measured

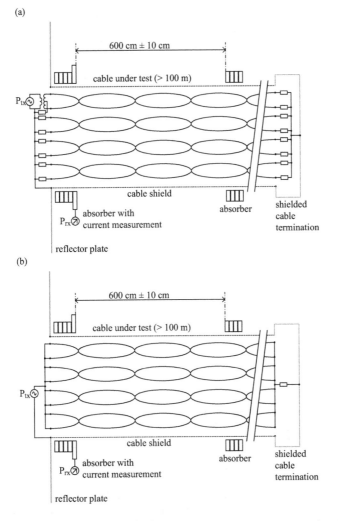

FIGURE 3.35 (a) Coupling and (b) shielding attenuation measurement of an STP-cable with absorbing clamp method (adapted from IEC 62153-4-5 2013).

separately by rewiring the cable under test as shown in Figure 3.35b. The unbalance attenuation (a_u) then can be calculated: $a_u = a_c - a_s$. Obviously, for coaxial and other unbalanced cables, the screening attenuation equals the coupling attenuation: $a_c = a_s$, $a_u = 0$.

The drawbacks of the absorbing clamp method are that they still require a relatively large space and do not exclude external effects unless a shielded room is used. In the *triaxial setup* (IEC 62153-4-4 2015; IEC 62153-4-9 2016; Halme 2013), shown in Figure 3.36a and b, these problems do not occur. The setup is comparable to the triaxial transfer impedance measurement shown in Figure 3.17, however, now the cables (and thus the setup) have to be electrically long. In this case the generator is feeding the inner circuit, because it is easier to match, optionally balance, and terminate it with its characteristic impedance leading to reflection free wave propagation in

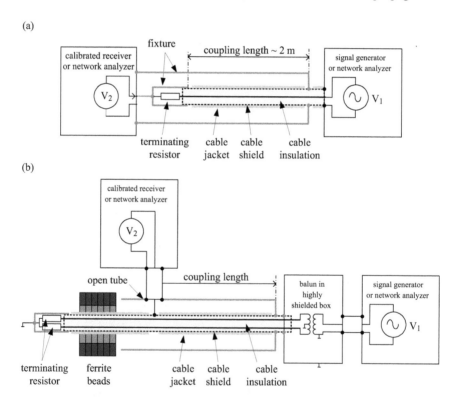

FIGURE 3.36 Triaxial setup for measuring the (a) shielding attenuation of cables and (b) the coupling attenuation of balanced cables (adapted from IEC 62153-4-4 2015 and IEC 62153-9 2018).

the excitation circuit. The outer circuit remains shorted at one end and the power leaked from the inner circuit is received at the other end.

If the setup of Figure 3.36 is used for qualifying connectors and cable assemblies the results for high-quality connectors are obscured by the connected cable. Therefore, the feeding cable can be placed in a tube as shown in Figure 3.37. This is called the *tube-in-tube* method (IEC 62153-4-7 2006)

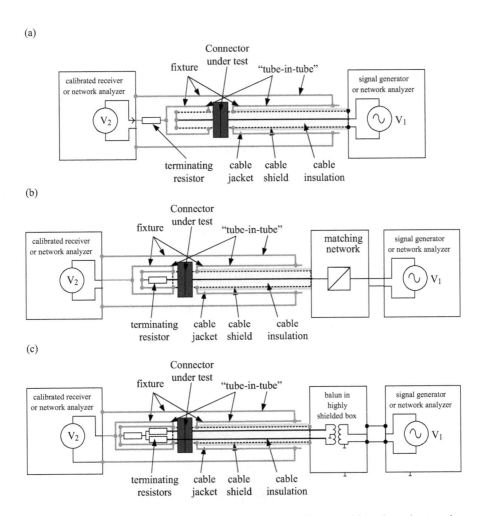

FIGURE 3.37 For characterizing connectors and cable assemblies the tube-in-tube method can be used, which nullifies the influence of the connection cable: (a) transfer impedance; (b) shielding attenuation; (c) coupling attenuation.

and can be used for measuring the transfer impedance at low frequencies as well.

3.4.6 Optical braid coverage

In the early days of EMC it was expected that optical translucency of cable shield would be a reasonable estimate of the EMC performance of a cable. Therefore, the term *optical (braid) coverage (OC)* came in use to characterize shields. OC is defined as follows (ANSI/SCTE 51):

$$OC = (2F - F^2) \times 100\%$$

$$F = \frac{N \cdot P \cdot D}{\sin(\alpha)} \tag{3.35}$$

$$\alpha = \tan^{-1}\left[2\pi(D + 2d)\left(\frac{P}{C}\right)\right]$$

where

OC is the optical (braid) coverage.

α is the Braid angle (radians) – the angle formed by the carriers with the longitudinal axis of the cable (refer to the illustration).

D is the diameter under the braid (inches).

C is the number of carriers – the number of groups of individual braid wires (ends), usually 16 for most cable telecommunications braided cables (refer to the illustration).

D is the Braid strand diameter (inches).

P is the picks per inch – the number of carrier crossing points per longitudinal inch (refer to the illustration).

N is the number of individual wires (ends) in each carrier.

See also Figure 3.38 for clarification of these parameters.

If one compares a shield with one large hole to a shield with four smaller holes such that the total open surface remains

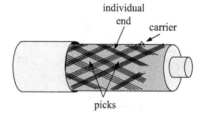

FIGURE 3.38   Parameters used to define cable braid coverage.

Table 3.2 Optical braid coverage and transfer impedance of various cables

Cable	Optical braid coverage	Transfer impedance	Reference
YmVkas	91%	130 mΩ/m	Authors
RG-58/U	96%	36 mΩ/m	Draka cables
RG-59/U	95%	36 mΩ/m	Draka cables
RG-174A/U	87%	45 mΩ/m	Miller and Vance cables
RG-213U	96%	8 mΩ/m	Cavel cables
RG-214/U	95%	< 0.9 mΩ/m	Cavel cables

equal the optical coverage will be equal. The EMC performance of the shield with the big hole, however, will be much worse (Vance 1974). This is confirmed by Table 3.2 where no correlation can be seen between the listed optical coverage and the transfer impedance. The RG-58/U and RG-213 have identical optical coverage but differ in a factor of 4.5 in transfer impedance. Similarly, the YmVkas cable has an optical coverage which is only 4% smaller than that of the RG-59/U, yet its transfer impedance is a factor of 3.6 higher.

3.5 Overview of standardized test methods

Over the past decades a variety of test methods has been developed. A small overview of standardized methods is given in Table 3.3 (IEC TS 62153-4-1 2014).

Table 3.3 Standardized methods to characterize cables, connectors, and assemblies (IEC TS 62153-4-1 2014)

Parameter	Method	Standard	Frequency range
Transfer impedance	Triaxial	IEC 62153-4-3 IEC 50289-1-6 IEC 61196-1	10 kHz–30 MHz
Transfer impedance	Matched T triaxial	IEC 60169-1-3	10 kHz–10 GHz
Transfer impedance	Line injection (time domain)	IEC 60096-4-1	1 kHz–80 MHz
Transfer impedance	Line injection (frequency domain)	IEC 62153-4-6	10 kHz–3 GHz

(Continued)

Table 3.3 (*Continued*) Standardized methods to characterize cables, connectors, and assemblies (IEC TS 62153-4-1 2014)

Parameter	Method	Standard	Frequency range
Shielding attenuation Coupling attenuation	Open screening attenuation test method	IEC 62153-4-5	30 MHz–1 GHz
Shielding attenuation	Reverberation chamber method	IEC 61726	300 MHz– 40 GHz
Shielding attenuation	Shielded screening attenuation test method	IEC 62153-4-4	10 kHz–3 GHz
Coupling attenuation of balanced cables	Current clamp injection	IEC 62153-4-2	50 MHz–1 GHz
Coupling attenuation of balanced cables	Coupling attenuation of screened balanced cables, triaxial method	IEC 62153-4-9	DC to 1 GHz
Transfer impedance Shielding attenuation Coupling attenuation	Tube-in-tube method	IEC 62153-4-7	DC to 3 GHz

CHAPTER FOUR

Protection of cabling and wiring

Cabling and wiring often form the coupling path between the source of electromagnetic interference and the victim who is susceptible against this interference. To reduce the coupling of and between cabling for internal and external fields the use of parallel earthing conductors (PECs) is recommended (IEC 61000-5-2 1997). Figure 4.1 shows an example where a measurement lead is connected to a sensor in a gas-insulated switch in the high-voltage grid (Deursen 1995). In the original situation, the sensor was connected with a single wire above the ground plane. During operation of the high-voltage switch, a surge of 15 kV was measured on the shorted wire. Adding a cable screen (by replacing the wire with an YMvKas cable), as discussed in Chapter 3, lowered the surge voltage with a factor of 50 to 282 V. When a cable tray was added to the installation and grounded at both ends, the voltage dropped with yet a factor of 100 to 2.4 V.

This chapter discusses tubular structures and nonmagnetic (aluminum) cable trays. Chapter 5 will discuss magnetic (steel) cable management systems, including nonlinear effects during power short circuit and lightning strokes.

4.1 Tubular structures

The *transfer impedance* concept (see Section 3.4.1) can be extended to characterize the Electromagnetic Compatibility (EMC) performance of PECs. In its definition (compare

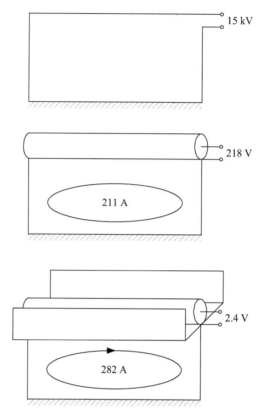

FIGURE 4.1 Surge voltage was reduced by a factor of 5000 by introducing a shielded cable and a cable tray.

Figures 3.18 and 4.1) the cable shield then has to be replaced by the external earthing conductor.

4.1.1 **Fully closed tubes**

Fully closed tubes offer the highest protection. At low frequencies the common mode current is homogenously distributed over the circumference of the tube as shown in Figure 4.2a and the transfer impedance equals the DC resistance. At higher frequencies the *skin effect* occurs: the common mode current starts concentrating at the outer surface. Effectively the bulk of the current only flows into the skin of the tube as shown in Figure 4.2b. The *skin depth*, δ, depends, next to the frequency, on the material properties:

$$\delta = \sqrt{\frac{1}{\pi f \mu_0 \mu_r \sigma}} \tag{4.1}$$

(a)   (b)

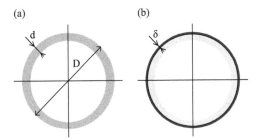

FIGURE 4.2   In coaxial cables, the cable shield is used as signal return conductor: (a) homogeneous current distribution when $d < \delta$; (b) skin effect when $d > \delta$.

where
  δ is the skin depth (m)
  f is the frequency (Hz)
  μ_0 is the magnetic permeability of vacuum ($4\pi \cdot \times 10^{-7}$) (Vs/Am)
  μ_r is the relative magnetic permeability
  σ is the conductivity (Ω^{-1} m^{-1})

The crossover frequency f_c between the low-frequency and high-frequency behaviors can be found by substituting $d = \delta$ in Eq. 3.1, where d is the wall thickness of the tube:

$$d = \delta = \sqrt{\frac{1}{\pi f_c \mu_0 \mu_r \sigma}} \Rightarrow f_c = \frac{1}{\pi \mu_0 \mu_r \sigma d^2} \qquad (4.2)$$

The transfer impedance Z_T per meter length can be approximated by Kaden (2006):

$$d < \delta \Leftrightarrow f_c < \frac{1}{\pi \mu_0 \mu_r \sigma d^2} : |Z_T| = R_{DC} = \frac{1}{\pi D d \sigma} \; (\Omega/\text{m}) \quad (4.3)$$

$$d > \delta \Leftrightarrow f_c > \frac{1}{\pi \mu_0 \mu_r \sigma d^2} : |Z_T|$$

$$= 2\sqrt{2} R_{DC} \frac{d}{\delta} e^{-d/\delta} \; (\Omega/\text{m}) \qquad (4.4)$$

where D is the cross section of the tube (m).

Figure 4.3 presents a measurement and a calculation on a copper tube. The correlation between theory and practice is very good. At low frequencies the transfer impedance is determined by the DC resistance of the copper tube. Above the crossover frequency, the transfer impedance starts to decrease exponentially.

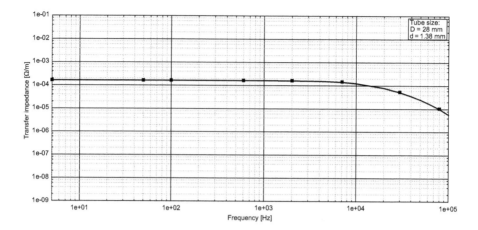

FIGURE 4.3 Transfer impedance per meter length of a copper tube. Dimensions are shown in the legend.

4.1.2 Shielded cables inside a tube

In practical applications, the tubular structure will not be used as signal return, but an additional conductor within the tube. In Figure 4.4, it is shown that this extra conductor leads to the formation of an *intermediate (IM) circuit*, next to the common mode (CM) and the differential mode (DM) circuits defined in Figure 3.2.

A CM current flowing over the tube causes a voltage to arise in the IM circuit:

$$V_{IM} = Z_{T,tube}I_{CM} \qquad (4.5)$$

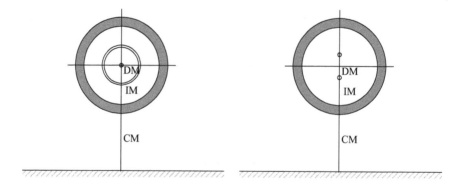

FIGURE 4.4 Application of multiple parallel conductors for the protection of the signal wire leads to the creation of an IM circuit.

This voltage leads to a current flow in the IM circuit:

$$I_{IM} = \frac{V_{IM}}{Z_{IM}} \tag{4.6}$$

where Z_{IM} is the impedance formed by the ground conductor and the tube. The current in the IM circuit causes a voltage in the DM circuit:

$$V_{DM} = Z_{T,cable}I_{IM} \tag{4.7}$$

Schematically, this is shown in Figure 4.5.

The overall transfer impedance can be calculated by multiplying the transfer impedances of the individual conductors and dividing by the impedance of the loop formed by the return wire and tube:

$$\left|Z_{T,total}\right| = \left|\frac{Z_{T,tube}Z_{T,cable}}{Z_{IM}}\right| \tag{4.8}$$

The transfer impedance of the RG-58/U itself has been measured with a triaxial setup (see Section 3.3.1) and is roughly up to 1 MHz equal to the DC resistance of its shield:

$$\left|Z_{T,RG58/U}\right| = R_{DC,RG58/U} = 13 \times 10^{-3} \ \left(\Omega/m\right) \tag{4.9}$$

The transfer impedance of the tubes already has been determined in Section 4.1.1. Slightly more complicated is the calculation of the impedance of the intermediate circuit. In principle, this is the sum of the resistance of the tube (R_{tube}), the resistance of the cable shield (R_{cable}), and the self-inductance (L_{IM}) of the loop formed by the cable shield and the tube; however, we have to take into account the *proximity effect*. Figure 4.6 shows that eddy currents (I_i) flow in the tube wall once the skin depth (δ) becomes smaller than the wall

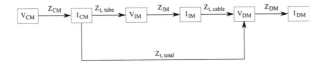

FIGURE 4.5 Scheme to calculate the overall transfer impedance of multiple parallel conductors.

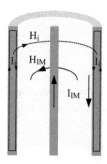

FIGURE 4.6 Half-section of a tube with a cable: at higher frequencies ($d > \delta$), the current through the tube causes a magnetic field in the tube wall, which in turn induces eddy currents in the tube wall.

thickness (d). On the inner surface the eddy currents have the same direction as the current in the IM circuit. At the outer surface they have the opposite direction. Since eddy currents form a loop, they generate a magnetic field (H_i) leading to a corresponding internal self-inductance (L_{tube}). In addition, the net current at the inner surface increases which leads to an increase of the resistive contribution of the tube to the IM circuit. Both the internal self-inductance and the resistive part depend on the frequency:

$$d < \delta \Leftrightarrow f < \frac{1}{\pi \mu_0 \mu_r \sigma d^2} : Z_{tube}(f) = R_{DC,tube} = \frac{1}{\pi D d \sigma} \quad (\Omega/\text{m})$$

(4.10)

$$d > \delta \Leftrightarrow f > \frac{1}{\pi \mu_0 \mu_r \sigma d^2} : R_{tube}(f) = \frac{1}{\pi D \delta \sigma} \quad (\Omega/\text{m})$$

$$L_{tube}(f) = \frac{1}{2\pi f} \frac{1}{\pi D \delta \sigma} \quad (\text{H/m})$$

(4.11)

$$|Z_{tube}(f)| = \sqrt{R_{tube}(f)^2 + [2\pi f L_{tube}(f)]^2} = \frac{\sqrt{2}}{\pi \delta \sigma} \quad (\Omega/\text{m})$$

The external self-inductance (L_{IM}) of the IM circuit depends on the magnetic field (H_{IM}) in the tube and can be approximated by the self-inductance formula for a coaxial cable:

$$L_{IM} = \frac{\mu_0}{2\pi} \ln\left(\frac{D_{tube}}{D_{cable}}\right) \quad (\text{H/m})$$

(4.12)

where D_{cable} is the diameter of the cable shield and D_{tube} is the diameter of the tube. In case of an RG-58/U cable, the diameter of the cable shield is 5 mm. With the diameter of the copper tube in the previous section ($D = 28$ mm), this leads to $L_{IM} = 1.5$ μH/m.

The total loop impedance can be approximated by

$$\left| Z_{IM}(f) \right| \approx \sqrt{\left[R_{tube}(f) + R_{cable} \right]^2 + \left[2\pi f L_{tube}(f) + 2\pi f L_{IM} \right]^2} \quad (\Omega/m) \tag{4.13}$$

Figure 4.7 shows the calculated impedance of the intermediate loop for an RG-58/U placed in the copper tube (dashed line with square markers) which has been discussed in the previous section. For low frequencies, Z_{IM} equals the DC resistance of the cable shield. At higher frequencies, the self-inductance becomes dominant. The (dashed line without markers) shows the calculated overall transfer impedance. These compare very well with the measured value (solid line with square markers).

It is apparent that the use of multiple PECs lowers the overall transfer impedance.

4.1.3 Slitted tubes

When a current is sent through a tube, it will generate a magnetic field around the tube, as has been demonstrated in Figure 3.3. As long as the tube is full closed around its circumference the

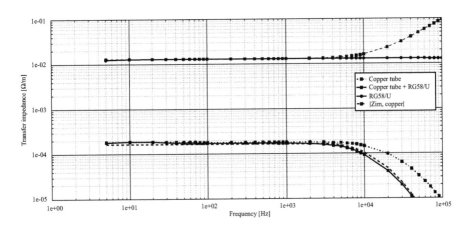

FIGURE 4.7 Overall transfer impedance of an RG-58/U cable in a copper. The dashed line without markers is the calculated transfer impedance and matches very well with the measurement (solid line with square markers).

magnetic field within the tube will remain zero. However, when a hole or slit is made in the tube, the magnetic field will penetrate through it. This is illustrated in Figure 4.8 for a copper a tube with a longitudinal slit.

The transfer impedance of the copper tube is at low frequencies nearly identical to the closed tube. The resistive part has slightly increased, because there is less conductive material due to the slit. This is confirmed by measurement results in Figure 4.9. Because the magnetic field leaks through the slit into the tube, some magnetic coupling (M_{slit}) will occur. This becomes notable at higher frequencies: the transfer impedance no longer exponentially decreases but starts to increase linear with frequency. In the higher frequency range, the transfer impedance can be written as

$$|Z_T| = 2\pi f M_{slit} \ (\Omega/m) \tag{4.14}$$

The amount of coupling with the magnetic field depends on the position of the inner conductor. When the inner wire is close to the slit the coupling is much higher than when the wire is remote from the slit and thus the transfer impedance depends on the position of the wire in the tube and is *no longer a mere property of the shield*. Figure 4.10 shows lines of constant magnetic coupling. These lines form circles all touching up on the slit. For sufficiently high frequencies ($d \gg \delta$), the coupling can be written as follows (Kaden 2006):

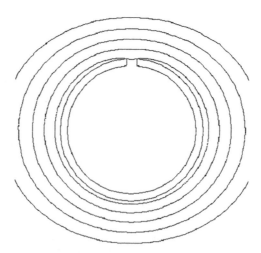

FIGURE 4.8 Magnetic field around and in a current carrying copper tube with a longitudinal slit.

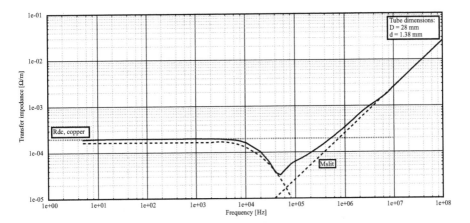

FIGURE 4.9 The transfer impedance of a copper tube and steel tube with a slit. The tubes have the same dimensions as the ones in Figure 4.3. The solid lines are measurements; the dashed lines are calculations for a closed tube and for $2\pi f M_{slit}$.

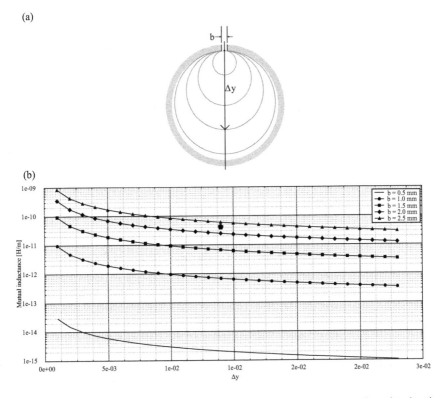

FIGURE 4.10 (a) Lines with constant mutual inductance in a tube with a slit. (b) The closer the inner wire to the slit and the wider the slit, the larger the coupling. The dot indicates the mutual inductance of the tube used in the tests.

$$d \gg \delta \Leftrightarrow f \gg \frac{1}{\pi\mu_0\mu_r\sigma d^2} : M_{slit}$$

$$= \mu_0 \left(\frac{1}{\Delta y}\right)\left(\frac{1}{\pi D}\right)\left(\frac{2b}{\pi}\right)^2 e^{-\pi\frac{d}{b}-2} \text{ (H/m)} \qquad (4.15)$$

where b is the width of the slit and Δy is the distance measured over the y-axis between the slit and the inner wire. Unfortunately, M_{slit} is highly dependent on the exact width of the slit and the precise position of the wire in the tube (as is obvious from Figure 4.10), which makes a reliable prediction difficult in practical situations.

The curves in Figure 4.9 are very similar to the cable transfer impedances shown in Figure 3.20. Solid tubes with holes or slits, therefore, are often used as calculation model for real cables.

4.1.4 **Commercial solutions**

Outside the United States, it is no common practice to route cables in conductive tubes. Nevertheless they can provide a very good solution. Alternatively there are commercial tubular products available. Some examples are given in Figure 4.11 and their measured transfer impedances are shown in Figure 4.12.

The transfer impedance of helical structures is very high. Figure 4.12 only shows the overall transfer impedance of such a construction with a cable inside. From this we can conclude that an additional structure only provides additional protection if it has lower or similar transfer impedance as the cable it has to protect. For meshed shields large differences are found, so it is important to inspect specifications carefully or perform additional measurements in case of doubt.

4.2 Nonmagnetic (aluminum) cable trays

In larger industrial and commercial environments, cable management systems are often used which are based on cable trays. When properly installed and connected, these trays drastically improve the EMC quality of the installation.

4.2.1 **Trays without a cover**

For demonstrating the behavior of nonmagnetic cable trays, the transfer impedance of a *plate*, a *shallow cable tray*, and a *deep tray* was investigated. The dimensions are shown in Figure 4.13. All three samples where made from an aluminum

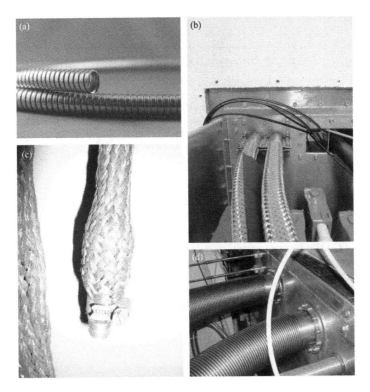

FIGURE 4.11 Some commercially available tubular products: (a) helical tube for mechanical protection; (b) braided tube for mechanical protection; (c) copper braid with RG-58/U; (d) bellow.

plate with a total width of 270 mm and a thickness of 1 mm. A shallow tray is defined as $h/w < 0.8$ and a deep tray as $h/w > 0.8$. The results for a measurement wire on the bottom (at a height of 2.5 mm) are shown in Figure 4.14.

4.2.1.1 Low-frequency behavior At low frequencies, the transfer impedance equals the DC resistance:

$$\left| Z_{T,tube}(f) \right| = R_{DC,tray} = \frac{1}{2(h+w)d\sigma} \ \left(\Omega/\text{m} \right) \qquad (4.16)$$

where
  f is the frequency.
  R_{DC} is the DC resistance of plate or tray.
  σ is the electrical conductivity (see Table 4.1).
  h is the height of the tray (= 0 for a plate).

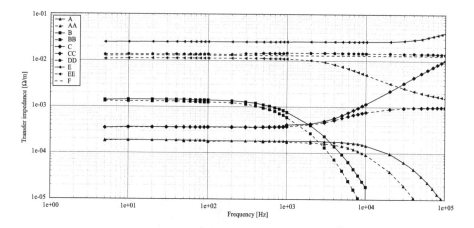

FIGURE 4.12 The transfer impedance of some commercially available tubular products: (A) copper tube; (AA) copper tube with RG-58/U; (B) steel tube; (BB) steel tube with RG-58/U; (C) copper braid; (CC) copper braid with RG-58/U; (DD) helical tube with RG-58/U; (E) braided tube for mechanical protection (Figure 4.11b); (EE) braided tube for mechanical protection with RG-58/U; (F) RG-58/U.

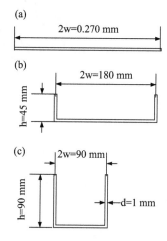

FIGURE 4.13 Cross section of (a) the investigated plate ($h/w = 0$), (b) the shallow tray ($h/w = 1/2$), and (c) the deep tray ($h/w = 2$).

> w is the half width of a tray.
> d is the material thickness.

> Obviously, there is no difference between the plate and the shallow and deep trays, because $h + w$ is kept identical. For all three geometries the transfer impedance reduces by 25% at relatively low frequencies. Eddy currents cause the current distribution within the plate and tray to become inhomogeneous: the current

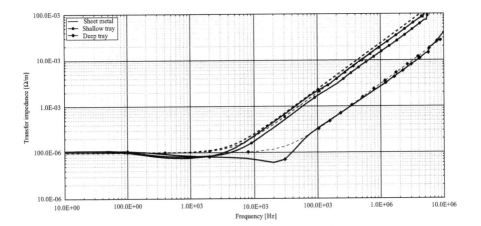

FIGURE 4.14 Transfer impedance of a plate, a shallow and a deep cable tray with a wire at the bottom in the middle. The solid lines represent measurements. The dotted lines are simplified approximations (see Eq. 4.20).

Table 4.1 Material properties of some commonly used materials

Metal	Conductivity σ	Relative permeability μ_r
Aluminum	$3.72 \times 10^{+7}$	1
Copper (CU-ETP)	$5.8 \times 10^{+7}$	1
Copper (CU-DHP)	$5.0 \times 10^{+7}$	1
Brass	$1.5 \times 10^{+7}$	1
Steel (tubular)	$6.5 \times 10^{+6}$	120 (small currents)
Steel (plates)	$8.5 \times 10^{+6}$	120 (small currents)
Zinc	$1.6 \times 10^{+7}$	1

Source: www.mcbboek.nl (entries in italics); other entries by measurement.

concentrates on the edges and reduces in the middle (see Figure 4.15). At higher frequencies, the skin depth becomes smaller than the thickness of the base material: $d > \delta$. Eddy currents then also flow over the thickness of the plate, as discussed for a tube in Section 4.1.1. In the transfer impedance curves of Figure 4.14, this is only visible for a deep tray.

4.2.1.2 High-frequency behavior At high frequencies the transfer impedance is determined by the mutual induction M_{tray} similar to the slitted tube in Section 4.1.3:

$$\left| Z_{T,tube}(f) \right| = 2\pi f M_{tray}(f) \left(\Omega/m \right) \qquad (4.17)$$

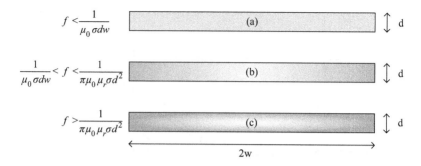

FIGURE 4.15 Eddy currents cause inhomogeneous current distribution in a metallic plate: (a) at low frequencies the current distribution is still homogeneous; (b) at middle frequencies the current in the middle section of the plate reduces and at the edges increases; (c) at high frequencies currents flow only near the surface of the plate. The largest current density is found in the corners.

Computer simulations and measurements have shown that the mutual inductance (M_{tray}) of plates and trays depend on the material properties, the frequency, and the dimensions. When the skin depth is very small compared with the thickness ($d > 5\delta$), the material properties can be ignored and the mutual inductance is solely determined by the dimensions and shape of the tray.

For small heights (Δy) of the measurement wire above a plate, the mutual inductance (M_{plate}) can be easily calculated (Kaden 2006):

$$d > 5\delta \Leftrightarrow f > \frac{25}{\pi\mu_0\mu_r\sigma d^2} : M_{plate} = \mu_0\Delta y\frac{1}{2\pi w} \text{ (H/m)} \quad (4.18)$$

The mutual inductance of a tray can be derived from the mutual inductance of the flat plate from which it has been formed (van Houten 1990):

$$M_{tray} = g \cdot M_{plate} = \mu_0\Delta y\frac{g}{2\pi(h + w)} \text{ (H/m)} \quad (4.19)$$

Computer simulations have been used to determine the shape factor for a variety of trays (see Table 4.2). Figure 4.16 shows that the shape factor rapidly decreases with increasing height/width ratio. Compared to cable shields, however, shallow cable trays still provide very good protection.

Figure 4.17 shows the mutual inductance as function of the position of the measurement wire within the tray. Values obtained with the formulas above apply to a surprisingly large

Table 4.2 Shape factor g of cable trays

h/w	g
0	1
1/4	0.99
1/2	0.84
3/4	0.65
1	0.51
1¼	0.37
1½	0.26
1¾	0.19
2	0.13
2½	0.063
3	0.031
5	0.0017

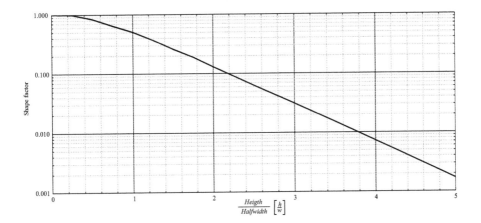

FIGURE 4.16 Shape factor g of cable trays.

area. Note that these figures are valid for any actual h and w, as long as the h/w ratio remains the same: if the height and width become 10 times larger, also the height of the measurement wire increase with a factor of 10. The best location to place cables is at the bottom in the corners.

4.2.1.3 Engineering equations For engineering purposes a rough estimate of the behavior of transfer impedance as function of frequency is sufficient in most situations. Such an estimate is given by

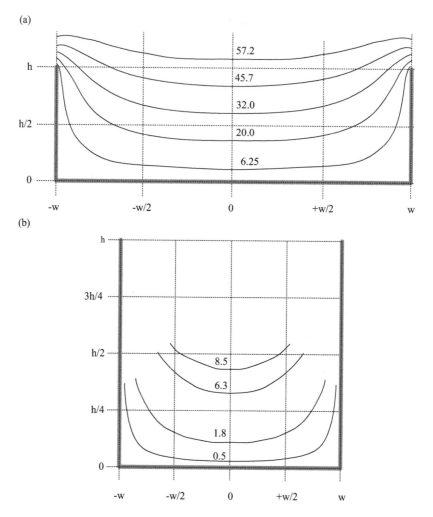

FIGURE 4.17 Mutual inductance as function of the position in (a) a shallow and (b) a deep tray. The mutual inductance depends only on the ratio of h/w and not on the actual values of h and w.

$$Z_{T,tray}(f) = \sqrt{R_{DC}^2 + (2\pi f M_{tray})^2}$$

$$= \sqrt{\left(\frac{1}{2\sigma d(h+w)}\right)^2 + \left(2\pi f \mu_0 \Delta y \frac{g}{2\pi(h+w)}\right)^2} \quad (\Omega/m)$$

$$(4.20)$$

These approximations compare very well with the measurements on experimental cable trays as shown in Figure 4.14.

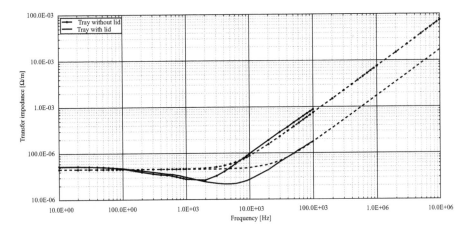

FIGURE 4.18 Transfer impedance of a commercially available aluminum tray with and without cover. The solid lines represent measurements. The dashed lines are simplified approximations.

Also for commercially available trays a valid quick estimate can be made. Figure 4.18 shows the results for such a tray with a width of 100 mm, a height of 68 mm, and a thickness of 2.5 mm. More accurate approximations are available from Helvoort (1995), but at the cost of complexity.

4.2.2 Trays with a cover

In most applications, aluminum is used for wall-mounted trays. These trays typically are covered with a cover. These covers are not in galvanic contact with the tray by themselves, because of the anodic treatment and/or painting. For safety reason, a grounding wire has to be connected to the cover, but this is still only a single-point connection and not permitting common mode current flow. Therefore, for low frequencies, there is no difference in transfer impedance for a tray with or without a cover. Nevertheless, the presence of the cover influences the mutual inductance and Figure 4.18 shows a factor of 4 reduction of the transfer impedance at higher frequencies. This factor can only be determined by measurement, because the width of the slit between cover and tray is hard to predict.

A disadvantage of aluminum trays is that it is hard to connect cable shields reliably to them.

4.2.3 Inter-connections

Normally, cabling has to span stretches which are longer than a single length of a cable tray. Therefore, multiple cable trays are used in series. For optimal EMC protection, it is important that they also are electrically connected. Each type of connection has its own localized, transfer impedance ($Z_{t,connection}$ in Ω), which can be approximated by

$$Z_{t,connection} = \sqrt{R_{connection}^2 + (2\pi f M_{connection})^2} \ (\Omega) \qquad (4.21)$$

where $R_{connection}$ is the resistance of the interconnection and $M_{connection}$ is the mutual induction of the interconnection.

Table 4.3 shows several different (electro-) mechanical interconnections and their parameters. Figure 4.19 shows the resulting transfer impedance. Note that the overall transfer impedance for can be found as follows:

$$|Z_{t,total}(f)| = |l \cdot Z_{T,tray}(f) + N \cdot Z_{t,connection}(f)| \ (\Omega) \qquad (4.22)$$

where l is the total length of the cable-carrying system (m) and N is the number of interconnections.

As can be observed from Figure 4.19, the best solution is to use a U-shaped interconnection with a large number of screws. The worst is a single-wire connection, which is the most typical situation in the shape of a ground-bonding wire. At higher frequencies, the interconnection with a bonding wire equals the transfer impedance of a 4.5 m tray.

4.2.4 Trays parallel to shielded cables

Figure 4.7 shows that it is beneficial to mount grounding structures in parallel. In Figure 4.20, this is demonstrated for a single-shielded cable (RG-58/U) in a cable tray.

Two positions are considered, one close to the bottom and one close to the opening. In both cases, when the cable shield is not connected at both ends, the same overall transfer impedance is found as for the unshielded cable. The difference between both positions is a factor of 10 at higher frequencies due to the increased inductive coupling for the cable higher in the tray.

4.2.4.1 Single-shielded cable

When the cable shields are connected at both ends the transfer impedance at lower frequencies hardly changes, because the resistivity of the cable shield is much larger than the resistivity of the cable tray. At high frequencies, however, the inductance of the loop formed by cable shield and tray flattens the transfer impedance curve, which can be explained by examining the equation for the total transfer impedance (see also Section 4.1.3):

$$|Z_{T,total}| = \left| \frac{Z_{T,tray} Z_{T,cable}}{Z_{IM}} \right| \qquad (4.23)$$

Table 4.3 Local transfer impedance of cable tray interconnections

			$R_{connection}$ [mΩ]	$M_{connection}$ [nH]
A	Number of ground wires	1	0.4	24
	Length ground wire	20 mm		
	Thickness ground wire	2.5 mm²		
	Gap between trays	1 mm		
B	Number of ground wires	2	0.2	8
	Length ground wire	20 mm		
	Thickness ground wire	2.5 mm²		
	Gap between trays	1 mm		
C	Number of ground wires	2	0.2	5
	Length ground wire	20 mm		
	Thickness ground wire	2.5 mm²		
	Gap between trays	1 mm		
D	Number of strips	2	0.1	2
	Number of screws	4		
	Strip length	20 mm		
	Strip height	90 mm		
	Strip thickness	1 mm		
	Gap between trays	1 mm		
E	Number of strips	2	0.5	1
	Number of screws	12		
	Strip length	20 mm		
	Strip height	90 mm		
	Strip thickness	1 mm		
	Gap between trays	1 mm		
F	U-shape	1	0.4	0.3
	Number of screws	4		
	Strip length	20 mm		
	Strip height and width	90 mm		
	Strip thickness	1 mm		
	Gap between trays	1 mm		
G	U-shape	1	0.2	0.2
	Number of screws	12		
	Strip length	20 mm		
	Strip height and width	90 mm		
	Strip thickness	1 mm		
	Gap between trays	1 mm		

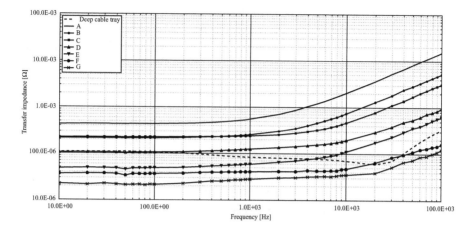

FIGURE 4.19 Local transfer impedance of cable tray interconnections (see Table 4.3). The dashed line indicates the transfer impedance of a 1-m-long deep tray.

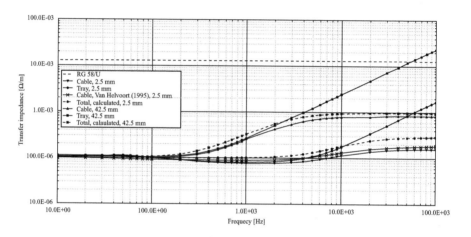

FIGURE 4.20 Overall transfer impedance of shielded cables in a tray.

The self-inductance L_{IM} of the cable shield in the tray can be approximated by the self-inductance of a wire above a ground plane:

$$L_{IM} = \frac{\mu_0}{2\pi} \ln\left(\frac{4\Delta y}{D}\right) \text{ (H/m)} \tag{4.24}$$

where Δy is the height of the cable in the tray (m) and D is the diameter of the cable (5×10^{-3} m for RG-58/U).

The contribution of the tray resistance to the intermediate (IM) loop is negligible with respect to the resistance of the cable shield and therefore can be neglected. With $Z_{T,RG\text{-}58/U} = R_{DC,RG\text{-}58/U} = 13 \times 10^{-3}$ (Ω/m), the total impedance per meter of the IM loop then becomes

$$|Z_{IM}(f)| \approx \sqrt{R_{DC,RG\text{-}58/U}^2 + (2\pi f L_{IM})^2} \quad (\Omega/m) \qquad (4.25)$$

The outcome of these approximations is shown with dashed lines in Figure 4.20 and compared well with the measurements. For the cable at the bottom, the transfer impedance is a little overestimated. For higher frequencies (above 100 kHz), the difference, however, reduces (Helvoort 1995).

4.2.4.2 Multiple cables   Normally, multiple cables are routed in a single cable tray. These cables shield each other form external disturbance. Effectively the cables high in the trays act similar to a cover for the cables lower in the tray. This is illustrated in Figure 4.21: most current runs via the upper cables.

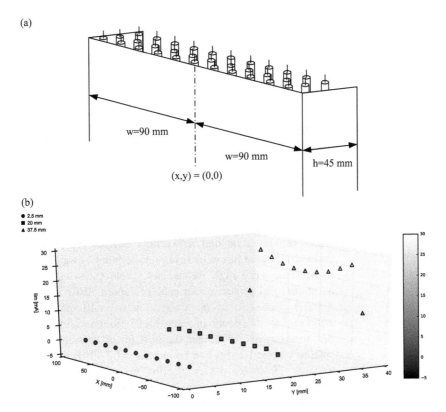

(a)

w=90 mm

w=90 mm

h=45 mm

(x,y) = (0,0)

(b)

● 2.5 mm
■ 20 mm
▲ 37.5 mm

FIGURE 4.21   Shielded cables higher in the tray protect the cables at the bottom of the tray: (a) setup; (b) simulation results. The total common mode current is 1 A.

The equations and graphs for a single cable provided earlier, therefore, can be seen as worst case approximations.

4.2.4.3 Cable separation   For both very sensitive and very disturbing cables it makes sense to place them at the bottom of the tray. Care, however, must be taken that these two types of cables are not mixed to prevent cross-talk. For separation most suppliers provide specific accessories. A full metal plate should be preferred. If good galvanic contact with the bottom is guaranteed effectively two deeper (and narrower) trays are created. Alternatively, sensitive and disturbing cables can be separated by using different cable trays for them.

4.2.5 Shielding effectiveness

The transfer impedance is a very effective method to characterize the EMC performance of cable trays. Nevertheless, the concept is only valid as long as the cross sections are small compared to the wavelength. The upper frequency limit (f_{SE}) is given by

$$h < 2w : f_{SE} = \frac{3 \times 10^8}{8w} \text{ (Hz)}$$

$$h > 2w : f_{SE} = \frac{3 \times 10^8}{4h} \text{ (Hz)}$$

(4.26)

The term 3×10^8 in these expressions is the speed of light in vacuum, h is the height of the tray, and w is half of its width. For the trays discussed in this chapter, the crossover frequency is between 300 MHz (a flat plate) and 1 GHz (the deep tray). Above this frequency shielding effectiveness (see also Section 3.4.4) should be used for characterization. In industrial environments, this is in particular importance due to the abundance of wireless communication and networking devices. In aeronautics and marine radar has to be taken into account as well.

At these very high frequencies, cable trays still provide additional EM protection for cables (Kapora 2010): the shielding effectiveness of an open tray is between 10 and 30 dB for all positions as shown in Figure 4.22. Remarkably, the corner positions are no longer the best for all frequencies. The shielding effectiveness can be increased by 30 dB by applying a cover. However, the connection between tray and cover becomes critical. To guarantee the behavior over the full frequency range, the cover has to be in full contact over the full length. In a temporary setup, this can be obtained by copper tape (Kapora 2010).

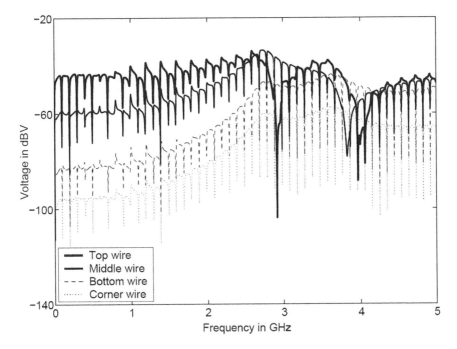

FIGURE 4.22 At very high frequencies, above f_{SE}, the corner and bottom position offer less protection than at lower frequencies (adapted from Kapora (2010)).

Special EMI/RFI (electromagnetic interference/radio frequency interference) shielding cable trays are also available commercially, a shielding effectiveness of 60 dB at 1 GHz has been reported (Chalfant). Strict installation instructions, in line with our previous findings, have to be met:

- Full mechanical release of tray interconnections
- Gap between sections, fittings, and cover should be kept as small as possible
- Clean contact areas before assembly
- Fill gaps and slits with tape or conductive filling
- Proper fixation of cover

CHAPTER FIVE

Protection of cabling and wiring with magnetic parallel earthing conductors

Parallel earthing conductors are very useful to reduce the coupling between cabling and their environment. Nonmagnetic structures have been discussed in detail in Chapter 4. The behavior of magnetic parallel earthing conductors is more complicated, but very important since most cable management systems are made from steel. Magnetic materials have an impact on the mutual inductance and can introduce nonlinear effects when the disturbance currents become large. This chapter discusses ferromagnetic tubular structures, ferromagnetic cable management systems and related nonlinear effects in case of short circuit and lightning currents.

5.1 Magnetic tubular structures

For small currents, the *transfer impedance* concept as discussed for tubular structures in Section 4.1 remains valid for characterization of magnetic structures.

5.1.1 Fully closed tubes

The transfer impedance behavior of fully closed, magnetic, tubes is identical to the nonmagnetic tubes. At low frequencies the common mode current is homogenously distributed over the circumference of the tube (as shown in Figure 4.2a)

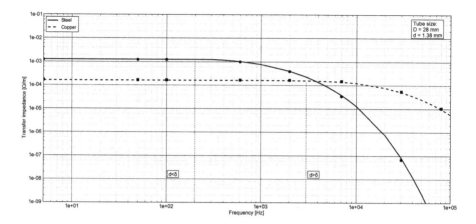

FIGURE 5.1   Transfer impedance of a copper tube (from Figure 4.3), and steel tube with identical dimensions. The vertical dotted line indicate the crossover frequencies where $d = \delta$ for both tubes.

and the transfer impedance equals the DC resistance. At higher frequencies the *skin effect* occurs: the common mode current starts concentrating at the outer surface (as shown in Figure 4.2b). Since the *skin depth, δ*, depends, on the material properties, this effect already occurs at lower frequencies (see Eq. 4.2). Figure 5.1 compares two tubes with identical dimensions: one copper and one steel. At low frequencies, the transfer impedance of the copper tube is lowest, because its DC resistance is lowest. However, the crossover frequency of the steel tube is smaller; therefore, the onset of the exponential decay starts earlier due to the magnetic properties of steel ($\mu \gg 1$). For closed structures steel is better than copper or other nonmagnetic metals. This conclusion is also valid for noncylindrical structures.

5.1.2 Shielded cables inside a tube

The approximations derived in Section 4.1.2 for determining the overall transfer impedance of a shielded cable in a tubular structure take magnetic materials into account. Figure 5.2 shows the calculated impedance of the intermediate loop for an RG-58/U placed in the copper tube and steel tube. For low frequencies, Z_{IM} equals the DC resistance of the cable shield. At higher frequencies, the self-inductance becomes dominant. The dashed lines show the calculated overall transfer impedance. These compare very well with the measured values (solid lines).

5.1.3 Slitted tubes

The transfer impedance of the slitted copper tube was found to be very similar to the closed copper tube. As shown in

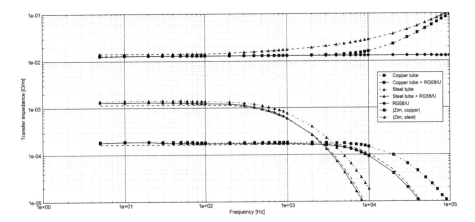

FIGURE 5.2 Overall transfer impedance of an RG-58/U cable in a copper and steel tube. The dotted lines show the transfer impedance of the individual tubes and cable. The dashed lines are calculations.

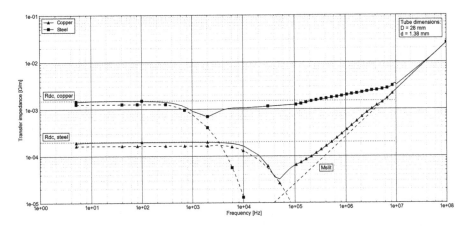

FIGURE 5.3 The transfer impedance of a copper tube and steel tube with a slit. The tubes have the same dimensions as the ones in Figure 8.1. The solid lines are measurements; the dashed lines are calculations for a closed tube and for $2\pi f M_{slit}$.

Figure 5.3 this is not the case for the magnetic steel tube. Over a very broad frequency range the performance of the steel tube is worse than of the copper tube due to the increased field penetration caused by the magnetic properties of steel as illustrated in Figure 5.4. *The protection of open magnetic parallel earthing structures is therefore less than their geometrical identical nonmagnetic counterparts.* At higher frequencies (in the MHz range), the transfer impedance becomes independent of the material properties and is only determined by the shape.

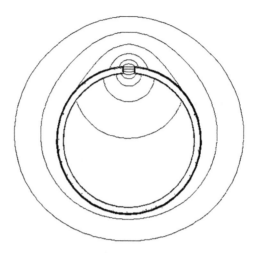

FIGURE 5.4 Magnetic field around and in a current carrying tube with a longitudinal slit.

There is no longer a difference between steel, copper, and any other conductive material.

5.2 Magnetic (steel) cable trays

Most industrial cable management systems are not made from aluminum, but from a sheet of steel. Stainless steel is also used under special circumstances and may be magnetic or non-magnetic. This can be easily checked with a small magnet. If the magnet sticks, the material is magnetic.

5.2.1 Solid trays without a cover

Magnetic field lines tend to follow the contours of the magnetic material; therefore, the behavior for open cable trays will be worse compared to similarly shaped trays from nonmagnetic materials. This is illustrated in Figure 5.5 where we compare steel trays with the aluminum trays discussed in Section 4.2: a *plate*, a *shallow cable tray*, and a *deep tray* were investigated. All three samples where made from a steel sheet (steel-37) with a total width of 270 mm and a thickness of 1 mm. A shallow tray is defined as $h/w < 0.8$ and a deep tray as, $h/w > 0.8$, where h is the height and w is half the width.

5.2.1.1 Low-frequency behavior   At low frequencies, i.e., when the skin depth is larger than the thickness ($d < \delta$), the transfer impedance equals the DC resistance:

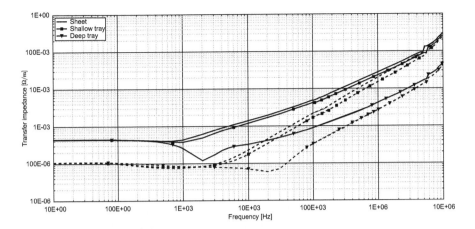

FIGURE 5.5   Transfer impedance of a steel plate sheet and a shallow and a deep cable tray with a wire at the bottom (solid lines) compared to the aluminum version (dashed lines) from Figure 4.13.

$$d > \delta \Leftrightarrow f > \frac{1}{\pi\mu_0\mu_r\sigma d^2} : \left|Z_{T,tray}(f)\right| = R_{DC,tray}$$

$$= \frac{1}{2(h+w)d\sigma} \ (\Omega/m) \qquad (5.1)$$

For a tray with a thickness of 0.8 mm, the low-frequency behavior is valid until 390 Hz, and with a thickness of 1.0 mm, it is valid until 250 Hz. In Figure 5.5, the transfer impedance of trays with a thickness of 1 mm is shown, and as predicted, DC behavior is shown until 250 Hz.

For a frequency higher than 250 Hz, there is a significant difference between shallow trays (including plates) and deep trays. For plates and shallow trays the transfer impedance increases due to the increase of the tray resistance caused by the skin effect. A deep cable tray behaves more similar to a slitted tube. Due to the fact that the field strength on the outside is much larger than on the inside the resistance dips between 250 and 6000 Hz. The value of 6000 Hz equals $d > \delta/5$. This latter formula has been validated by a large number of simulations (Helvoort 1995).

5.2.1.2 High-frequency behavior   Equivalent to the behavior of slitted tubes, the transfer impedance of cable trays becomes independent of the material parameters at high frequencies. The expression (Eq. 4.19) derived for nonmagnetic cable trays becomes valid when the frequency f meets both the following conditions:

$$\left. \begin{array}{c} f > \dfrac{\mu_r}{2\pi\mu_0\sigma(\Delta y)^2} \\[4mm] f > \dfrac{\mu_r}{2\pi\mu_0\sigma d\sqrt{h^2+w^2}} \dfrac{\pi-\ln(1+(h/w)^2)}{2g} \end{array} \right\} \Rightarrow M_{tray}$$

$$= g \cdot M_{plate} = \mu_0 \Delta y \dfrac{g}{2\pi(h+w)} \quad (\text{H/m}) \qquad (5.2)$$

The shape factor g has been given in Table 4.2. For wires high in the tray, the first condition for the frequency is dominant. For wires located near the bottom of the tray the second frequency is dominant. For shallow trays ($h/w < 0.8$), the second frequency condition can be simplified:

$$f > \dfrac{\mu_r}{2\pi\mu_0\sigma d\sqrt{h^2+w^2}} \dfrac{\pi}{2g} \qquad (5.3)$$

For typical dimensions, these frequencies are in the upper kHz range or lower MHz range.

5.2.1.3 **Engineering equations**   For engineering purposes, a rough estimate of the behavior of transfer impedance as function of frequency is sufficient in most situations. Only for shallow trays ($h/w < 0.8$), a sufficient expression has been found:

$$\left| Z_{T,tray,shallow,magnetic}(f) \right| = \sqrt{\left[R(f)\right]^2 + \left[2\pi f L(f) + M_{tray}\right]^2} \quad (5.4)$$

This approximation resembles Eq. 4.20, when the DC resistance of the nonmagnetic tray is replaced by the frequency dependent internal impedance of the steel tray. This internal impedance is split into a resistive part $R(f)$ and an internal self-inductance $L(f)$:

$$d < 2\delta \Leftrightarrow f < \dfrac{2}{\pi\mu_0\mu_r\sigma d^2}$$

$$\Rightarrow \left\{ \begin{array}{c} R(f) = R_{DC} = \dfrac{1}{2\sigma d(h+w)} \quad (\Omega/\text{m}) \\[4mm] L(f) = 0 \quad (\text{H/m}) \end{array} \right. \qquad (5.5)$$

$$d > 2\delta \Leftrightarrow f > \frac{2}{\pi\mu_0\mu_r\sigma d^2}$$

$$\Rightarrow \begin{cases} R(f) = \dfrac{1}{\sigma\delta} \dfrac{g}{2\pi(h+w)} \quad (\Omega/\text{m}) \\[3mm] L(f) = \dfrac{1}{2\pi f} \dfrac{1}{\sigma\delta} \dfrac{g}{2\pi(h+w)} \quad (\text{H/m}) \end{cases} \tag{5.6}$$

Substituting Eqs. 5.5 and 5.6 into Eq. 5.4 leads to

$$d < 2\delta \Leftrightarrow f < \frac{2}{\pi\mu_0\mu_r\sigma d^2} \Rightarrow \left| Z_{T,\text{tray,shallow,magnetic}}(f) \right|$$

$$= \sqrt{\left(\frac{1}{2\sigma d(h+w)} \right)^2 + \left(2\pi f\mu_0\Delta y \frac{g}{2\pi(h+w)} \right)^2} \quad (\Omega/\text{m}) \tag{5.7}$$

$$d > 2\delta \Leftrightarrow f > \frac{2}{\pi\mu_0\mu_r\sigma d^2} \Rightarrow \left| Z_{T,\text{tray,shallow,magnetic}}(f) \right|$$

$$= \frac{g}{2\pi(h+w)} \sqrt{\frac{1}{(\sigma\delta)^2} + \left(\frac{1}{\sigma\delta} + 2\pi f\mu_0\Delta y \right)^2} \quad (\Omega/\text{m}) \tag{5.8}$$

Figure 5.6 proves the usability of the engineering expressions for shallow trays. Even for deep trays a fair estimate can be made with exception of the frequency range between 1 and 10 kHz.

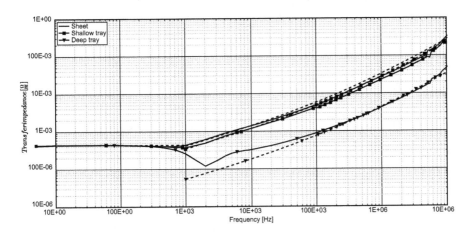

FIGURE 5.6 Transfer impedance of a steel plate and a shallow and a deep cable tray with a wire at the bottom (solid lines) approximated by engineering expressions (dotted lines).

5.2.2 **Commercially available cable routing systems**

Commercially, a variety of cable trays are available, varying from solid-walled trays to very open cable ladders or cable baskets (wire mesh trays). Several examples are listed in Table 5.1.

5.2.2.1 ***Solid trays (C1)***   Cable tray C1 is a solid tray without perforations and ports. As is shown in Figure 5.7, its transfer impedance can be reasonably well approximated using Eq. 5.8. The actual transfer impedance slightly decreased because C1 has upper edges that are bent to the inside of the tray. This leads to a 13% underestimation of the plate width in the engineering expression, therefore, both the calculated resistance and the mutual inductance are overestimated (respectively by 13% and 30%). Though zinc has a higher conductivity the layer thickness (approximately 19 µm per side) is too small to lower the overall conductivity. However, it reduces $R(f)$ and $L(f)$, compared to the nonplated material, when the skin depth becomes smaller than the thickness of the tray. The worst case deviation with respect to the engineering expressions is 50%.

5.2.2.2 ***Perforated trays (C2, C3, and C7)***   Cable tray C2 is nearly identical to C1 (as shown in Figure 5.7) but contains some perforations in the side wall and bottom. The holes increase both the resistance and mutual inductance of the tray.

Table 5.1   Identification and description of commercially available cable trays (see also ssures 5.7–5.10)

Number	h/w	h (mm)	$2w$ (mm)	d (mm)	Description
C1	0.6	60	200	1	Zinc-plated steel cable tray without perforations
C2	0.6	60	200	1	Zinc-plated steel cable tray with some perforations in the side walls and bottom
C3	0.3	60	400	1	Zinc-plated steel cable tray with side perforations and corrugated bottom
C4	0.6	60	200	1.25	Zinc-plated steel cable ladder Distance between rungs: 200 mm
C5	0.3	60	400	1.25	Zinc-plated steel cable ladder Distance between rungs: 200 mm
C6	0.62	60	200	4.4	Cable basket (wire mesh cable tray) with 10 longitudinal steel wires (4 mm outer diameter)
C7	0.1	15	300	1.25	Zinc-plated steel cable tray with many transverse perforations
C8	1.36	68	100	2.5	Anodized aluminum tray (see Figure 7.18)

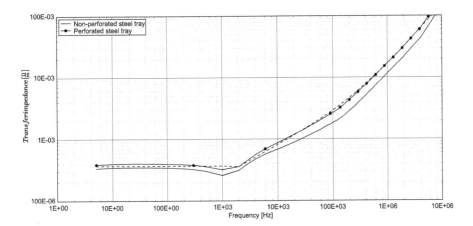

FIGURE 5.7 Transfer impedance of two zinc-plated steel cable tray with identical cross section, C1 without and C2 with perforations and ports. The solid lines represent measurements. The dotted line is calculated (Eq. 5.8).

Nevertheless, the maximum difference between C1 and C2 over the full frequency range is limited to 30%. Coincidentally, the extra leakage of the magnetic field through the holes "compensates" the field reduction caused by the bent upper edges which was not taken into account by Eq. 5.8.

A corrugated bottom profile is often used to increase the stiffness of wide cable trays. This will slightly increase the magnetic coupling as is shown in Figure 5.8. Nevertheless, in practical applications, the engineering expressions, while ignoring the corrugation, lead to acceptable results.

When many more holes are added, such as is the case for tray C7, it is no longer possible to predict the mutual induction (see Figure 5.8). Note that so much material has been removed that also the DC resistance is significantly higher. Nevertheless even this kind of trays offer significant protection compared to cable shields.

5.2.2.3 *Cable ladders (C4 and C5) and cable baskets (C6)* In addition to cable trays, much more open cable routing systems are employed, like cable ladders and cable baskets (also branded as steel wire cable trays). The transfer impedance is larger than of comparable trays, due to larger openness and reduction in metallic cross section (see Figure 5.9). Actually in cable ladders the rods add little to the protection of the cables and the sides can be considered parallel plates. Best protection is obtained when a cable is mounted close to one of the frame sides. The transfer impedance than can be estimated by entering the dimensions of a single plate in Eq. 5.8.

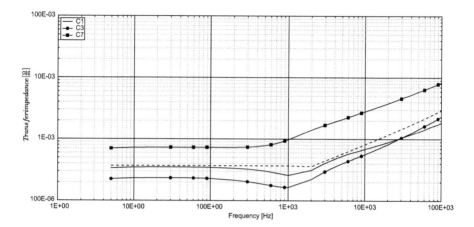

FIGURE 5.8 Transfer impedance of a zinc plated-steel cable tray with a corrugated bottom (C3) and with many perforations (C7). The solid lines represent measurements. The dotted line is calculated (Eq. 5.8). The dashed line shows the measured transfer impedance of C1.

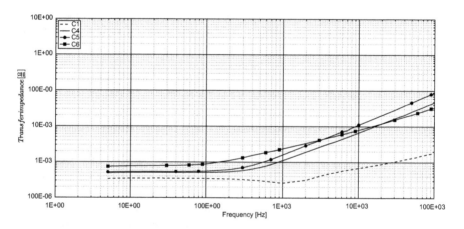

FIGURE 5.9 Transfer impedance of a zinc-plated steel cable ladders (C4 and C5) and a cable basket (C6). The solid lines represent measurements, with the measurement leading the middle of the structure. The dashed line shows the measured transfer impedance of C1.

For cable baskets, the engineering expressions are not valid. The best location of a cable is close to one of the longitudinal wires (Maciel 1993).

5.2.2.4 Trays with a cover (C1 and C2)   Some cable trays are available with a cover. Similar to the results of aluminum tray discussed in Section 4.2.2 the reduction is the strongest once the skin effect is present (see Figure 5.10). At frequencies above

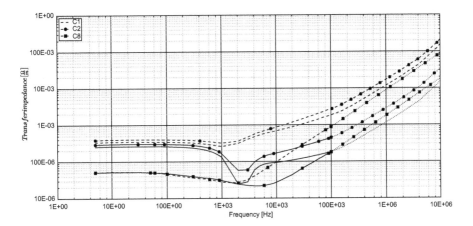

FIGURE 5.10   Transfer impedance of a zinc-plated steel cable tray with a cover (C1 and C2) compared to an aluminum tray (C8). The solid lines represent measurements. The dotted lines are calculations (Eq. 5.8). The dashed lines show the transfer impedance of the trays without cover.

10 kHz the reduction is a factor of 10 for the unperforated tray (C1) and a factor of 6 for the perforated tray (C2). This indicates that magnetic field lines rather close via the cover, than via the inside of the tray. The measured curves are similar to those of a steel tube with a slit (compare Figure 5.10 with Figure 5.3).

Measurements on the trays with the cover were only performed to 100 kHz, for higher frequencies the results have been extrapolated using Eq. 5.8 (indicated with the dotted lines). At low frequencies also some reduction in transfer impedance is found, because the zinc plating improves the electrical contact between the tray and its cover which reduces the overall DC resistance.

5.2.3 Trays parallel to shielded cables

The transfer impedance of a shielded cable in a nonmagnetic tray is given by Eq. 4.25. For most cable positions in a magnetic tray Eq. 4.25 remains valid, with the exception when the cable is routed close to the bottom. In the latter case, the internal impedance of the tray has to be taken into account for which only a solution exist for high frequencies (Helvoort 1995). Measurement and calculations are compared in Figure 5.11.

5.3 Construction steel

In situations where only one or a few cables have to be routed, it is not essential to use a cable tray to protect the cable, but

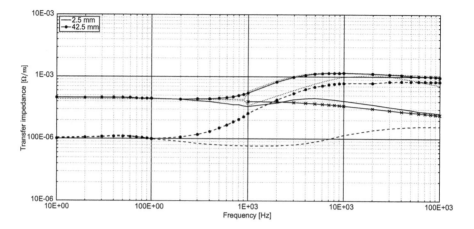

FIGURE 5.11 Transfer impedance of a shielded cable in the shallow steel cable tray from Figure 8.5. The solid lines represent measurements. The dotted lines are calculations (Eqs. 4.20, 4.25, and 5.8). The line with x-marks is taken from Helvoort (1995). The dashed lines show the transfer impedance of a cable in an identical aluminum tray.

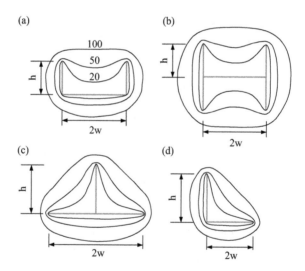

FIGURE 5.12 Construction elements can be used as alternative for a cable tray. The lines around the elements indicate the value of the mutual induction at the given position: the inner line is 20 nH/m, the middle line is 50 nH/m, and the outer line is 100 nH/m (Deursen 2000).

construction steel, readily available in the machine or building, can provide an excellent alternative. In Figure 5.12, several examples are given.

Flat strips have identical behavior as plates. U-shaped beams are comparable to cable trays. H- and I-beams act like two cable

Table 5.2 DC resistance and mutual inductance of some typical construction elements

Shape	Type	Resistance ($\mu\Omega$/m)	Mutual induction (nH/m)
Angle bar	$140 \times 140 \times 15$	25	0.3
Parallel flange channel	UNP 220	27	1.9
Universal beam	IPE 240	26	1.9
Universal column	HE 160-A	26	1.0

The measurement wire is at a height of 2.5 mm.

trays back to back. As long as the thickness d of the construction elements is small compared to the other dimensions, the mutual inductance can be expressed as:

$$M_{tray} = \mu_0 \Delta y \frac{2g}{\pi C} \text{ (H/m)} \tag{5.9}$$

where C is the length of the contour, i.e., the circumference, of the construction element. For a U-shaped cable tray, the circumference C equals $4(h + w)$. When substituted in Eq. 5.9 obviously Eq. 4.19 is found. Table 5.2 shows some examples of construction steel and their transfer impedance properties.

5.4 Nonlinear behavior in fault situations

When steel cable management systems are properly integrated in the grounding systems, the currents flowing over them will remain small under normal circumstances. However, in fault situations, like lightning stroke or major short circuits, large currents may arise. As steel shows nonlinear behavior, the transfer impedance concept must be applied with caution in these situations:

- The induced voltage may be higher or lower than expected.
- The induced voltage may contain lower and/or higher harmonics.

The magnetic properties of materials can be described with a "B–H curve." An example is given in Figure 5.13 for aluminum and steel. Since aluminum is linear, the relation between B and H is a straight line (i.e., $B = \mu_0 H$). The data for steel are

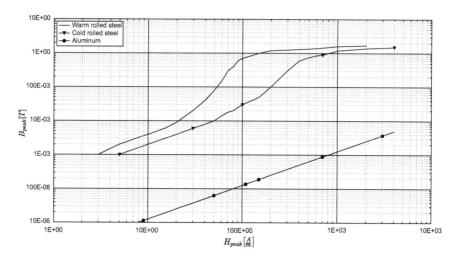

FIGURE 5.13 Static *B–H* curve for cold- and warm-rolled steel (Küpfmüller 1973) compared with an experimentally determined curve of the steel used in Section 5.2. The straight line depicts aluminum.

taken from Küpfmüller (1973). The behavior of steel depends on the production process. For the trays in Section 5.2, a *B–H*-curve was determined experimentally. At low values of *H*, till $H = 40\,\text{A/m}$, this measured curve can be approximated by a straight line through the origin: $B = \mu_0\mu_r H$. The relative permeability associated with this approximate line is called the *initial permeability* line and equals a value of $\mu_{r,initial} = 120$.

When the magnetic field increases above $40\,\text{A/m}$, the *B–H*-curve bends upward. The relative permeability μ_r increases up to 2500. The magnetic flux density does not increase any further when the magnetic field strength is increased, which is called *saturation*. Nevertheless, the distortion voltage will remain high at very high currents (40 kA) and high frequencies (1 MHz). Our hypothesis is that the maximum deviation of the actual peak voltage compared to the peak voltage predicted by a linear transfer impedance is:

$$\frac{V_{peak,actual}}{V_{peak,linear}} = \sqrt{\frac{\mu_{r,max}}{\mu_{r,initial}}} = \sqrt{\frac{2500}{120}} \approx 5 \qquad (5.10)$$

For alternating current or repetitive pulses, there is a second effect to be taken in to account, because magnetic materials have hysteresis. This means that the *B*-field caused by an increasing *H*-field differs from the *B*-field caused by a diminishing *H*-field. This is graphically depicted in Figure 5.14.

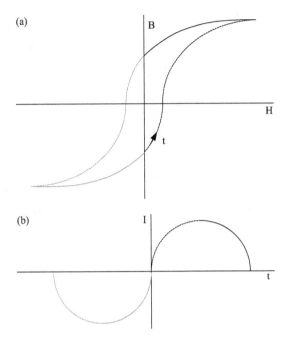

FIGURE 5.14   The upper graphs show an example of a dynamic *B–H* curve (hysteresis). The lower graph shows a sinusoidal excitation current. The graphical representation of the segments corresponds with each other.

5.4.1 Short circuits

5.4.1.1 Mains short circuit   In industrial installations, large 50 or 60 Hz currents may flow in the grounding structures during a short circuit. In the setup of Figure 5.15, the behavior of the disturbance voltage caused by large currents over a nonlinear steel tray is studied. For a 50 Hz current the results are shown in Figure 5.16, where the peak voltage is plotted against the peak current of a 50 Hz current. Based on the linear transfer imped-ance concept one predicts that the peak voltage will double if the current doubles (dotted line). For currents above 100 A the peak voltage becomes larger than predicted linearly.

Figure 5.17 shows the distortion of the DM voltage, which has become almost triangular instead of sinusoidal.

Fourier analysis describes an arbitrary wave shape as a sum of sinusoidal signals and the measured triangular voltage shape mainly consists of three frequencies: the base frequency at 50 Hz and the third and fifth harmonic at 150 and 250 Hz, respectively. The amplitude of the sinusoidal signal varies with frequency. In the given example the 50 Hz signal is the largest and the 250 Hz signal the smallest.

FIGURE 5.15 Experimental setup allowing large currents to flow over a steel grounding structure. The red sensor around the tray is a *Rogowski coil* measuring the total current through the tray.

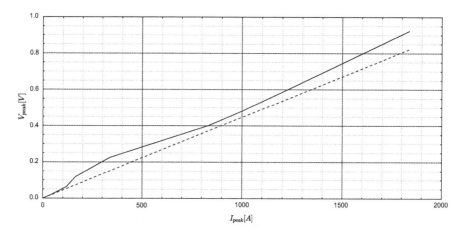

FIGURE 5.16 Peak disturbance voltage measured in a shallow tray, for a 50 Hz common mode current. The dashed line shows the linear prediction $Z_t I_{peak}$, which is only valid for small currents.

The nonlinear effect of steel introduces odd harmonics at 150, 250 Hz, and above that cannot be predicted by the transfer impedance concept. Note that the left (voltage) axis has been chosen such that the dashed line reflects the linear prediction, i.e. $Z_t I_{CM}$. In this particular case, the peak voltage is 1.5 times as high as predicted when using the transfer impedance. As we have seen in Figure 8.13 the nonlinearity related to an

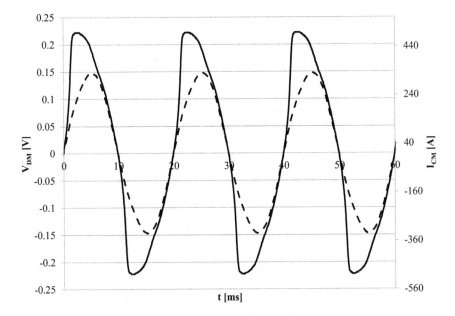

FIGURE 5.17 The nonlinear behavior of steel causes distortion of the disturbance voltage (solid line). This measurement was made with a shallow tray and a 50 Hz current dashed line. The curves were aligned at the zero crossings and the dashed line can be read as $Z_t I_{CM}$ on the left-hand scale.

increasing current is larger than with a declining curve; therefore, the distortion in the voltage is the largest during its onset.

For more scientific purposes the 50 Hz measurements have been repeated at 250 and 500 Hz. Again up to a current of 100 A_{peak} the behavior remains linear. The hypothesis for any shape tray made from regular plate steel is that behavior is linear up to a current of:

$$I_{linear} \leq 555(h + w) \text{ (A)} \qquad\qquad (5.11)$$

where h is the height of the tray and w is its half width. Neither confirmation nor objection for this hypothesis is known at the time of writing. Care must be taken that cold-rolled and warm-walled steel will behave differently.

5.4.1.2 *Short circuits in capacitor banks*   With the emergence of photovoltaic energy generation, DC storage in capacitor banks becomes more common place. In such installations, large damped sinusoidal currents at higher frequency may occur when a capacitor is incidentally shorted to the grounding

FIGURE 5.18   The discharge of a capacitor can cause large currents with high frequency.

system. In Figure 5.18, a high-voltage capacitor is added to the setup of Figure 5.15. The discharge current was varied between 9 and 40 kA at a frequency of 20 kHz and between 500 and 7 kA at a frequency of 1 MHz.

Figure 5.19 shows the measured discharge current and interference voltage. The scales are chosen such that the right-hand scale shows I_{CM} and the left-hand scale shows both V_{DM} and $Z_t I_{CM}$. Interestingly, the prediction of the peak voltage using $Z_t I_{CM}$ is very fair for high current magnitudes (i.e., from 0 to 100 µs). At less current (i.e., for $t > 100$ µs), the deviation is much larger and goes up to a factor 5. This matches Eq. 5.10.

5.4.2 **Lightning stroke**

Direct lightning strokes can cause currents in to the 100 kA range. Fortunately, in most cases cable trays are used within buildings and only indirect currents with a smaller amplitude will flow. These indirect currents were simulated by connecting a Marx surge generator, which normally is used for high-voltage component testing, to the setup of Figure 5.15. This resulted in the typical lightning pulse with a peak of 1 kA

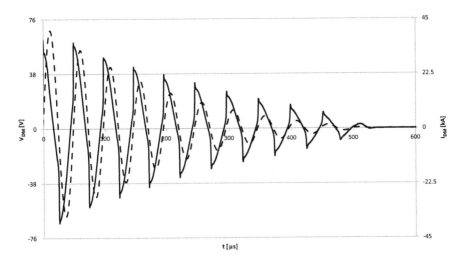

FIGURE 5.19 Disturbance voltage in a shallow steel tray caused by capacitor discharge. The frequency is 20 kHz.

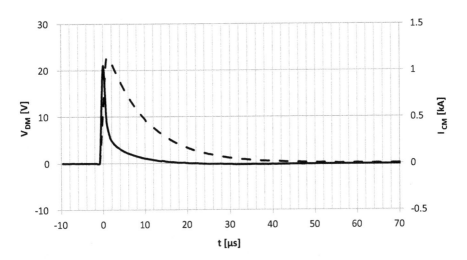

FIGURE 5.20 Lightning pulse commonly used in high-voltage testing. The dashed line shows the measured lightning current, and the solid line shows the disturbance voltage in a steel tray.

shown in Figure 5.20. Figure 5.21 shows the Fourier transform of the measured current and voltage waves. The spectrum of the lighting current (I_{CM}) has the largest content at low frequency. For a frequency, higher than 10 kHz, the amplitude drops off. Hence, the occupied frequency spectrum is roughly 1 MHz. The low-frequency components as well as the DC component

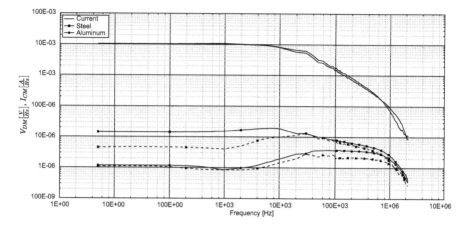

FIGURE 5.21 Frequency content of the lightning pulse. Dashed lines are calculated under assumption of linearity. For the aluminum tray measurement and calculation match well. For the steel tray only at higher frequencies the voltage is estimated correctly.

are the root cause for the fire hazard posed by lightning. The higher frequency components, however, cause the most damage to electronic circuits, because they are responsible for the sharp peak in the time domain (Figure 5.20). Conventional lightning protection measures focus on the reduction of the fire hazard and are insufficient for broadband protection of micro-electronics.

If we assume linear behavior of the cable tray used for additional protection, the voltage V_{DM} can be predicted with the transfer impedance:

$$V_{DM}(f) = Z_t(f)I_{CM}(f) \tag{5.12}$$

In case of aluminum trays, Figure 5.21 shows that the prediction and measurement match very well. In the case of a steel (nonlinear) tray, the predicted value is only 30% of the actual measured value at low frequencies. At higher frequencies, which we have shown to be more relevant for electromagnetic compatibility, they match very well.

5.4.3 Tubular structures

Tubular structures are commonly used in the Unites States and should be considered elsewhere as well for sensitive connections or connections in the open air. Figure 5.22 compares the performance of copper and steel tubes. Again, if steel is selected, nonlinear behavior can be expected. However, if the structure is completely closed around its circumference,

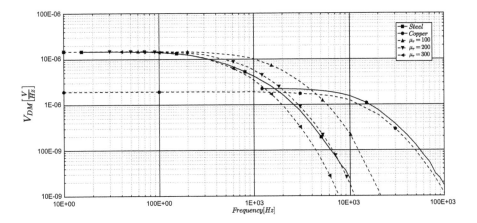

FIGURE 5.22 Frequency content of the disturbance voltage caused by a lightning current over a copper and a steel tube. Dashed lines indicate $Z_t(f)I_{CM}(f)$.

the actual disturbance voltage will be lower than the voltage predicted under assumption of linearity. This effect typically becomes visible at slightly higher frequencies (above 100 Hz).

Already in 1970, closed structures made from magnetic material were proposed by Ferber and Young (1970) and Merewether (1970) for defense and nuclear installations.

5.4.4 Conclusion

Only in extreme situations, such as shortening in large power circuits or lightning strikes, the assumption of linearity will underestimate the disturbance voltage. In the design phase, this can be dealt with by taking a safety factor 5 into account. Care must be taken that steel structures may cause a broadband disturbance voltage of which the spectrum is 13 times as broad as the current.

CHAPTER SIX

Barriers against conducted disturbances

In the previous chapters, it was shown how to minimize induced disturbance voltages in cables and wiring by protecting them with grounding structures, which are allowed to carry common mode (CM) currents. When passing from one *zone* to another (see Section 1.2), a *barrier* needs to be in place such that the constraints on conducted disturbances are met in both zones. Rerouting currents around a more susceptible zone may be required such that they will not flow over or near more sensitive equipment or electronics.

This chapter discusses grounding strip, back panel, full enclosures, connection of cable shields, connection of non-electrical conductors, connection of cable trays, use of ferrite, multipoint grounding of cables, buried shield wires, filters and overvoltage protection.

6.1 Partial and full enclosures

6.1.1 Ground strip Goedbloed (1990) recommends designers of electronic systems to connect all cables on one site of the printed circuit board (PCB). At that site, a short and broad grounding trace should connect all cable shields and other signal grounds. This concept should be extended to both conductive and nonconductive enclosures and cabinets (see Figure 6.1):

- All cables have to enter on the same site.
- Use a short, broad, strip to connect all grounds, i.e., *grounding rail* (in particular, for nonconductive housings, for conductive housings; see also Section 6.2).

FIGURE 6.1 Proper grounding of cables and shields inside a cabinet: (a) cabinet with ground rail; (b) cable clamp.

The illustration in Figure 6.2a shows two shielded cables which are connected to a small PCB. The shields are connected to each other via a trace on the board. Due to external influences, a CM current flows over both the shields and the PCB trace. This current will induce a voltage over the interruption in the second trace on the board, because both traces form a (planar) *transformer*. In a real world example, this interruption could be the high-ohmic input impedance of a sensitive amplifier. Already a small disturbance voltage at its input could lead to large deviations in the full system; therefore, the disturbance voltage should be kept as low as possible.

In the given example, a CM current of 1 A at a frequency of 10 MHz would lead to a disturbance voltage of 0.61 V. When the print layout is improved such that only a short trace is used to connect the cable shields (Figure 6.2b), this voltage drops with a factor of 10 to 62 mV.

6.1.2 **Back panel**

The coupling between the traces in Figure 6.2b can be further reduced by replacing the planar ground strip by a panel as shown in Figure 6.2c. Finite Element computer simulations indicate a further reduction by factor of 8 to a disturbance voltage of 7.4 mV. Alternatively, the *back panel* can be connected in parallel to the ground trace on the PCB with the same positive result. A PCB with a poor electromagnetic compatibility (EMC) design thus can be improved by adding external measures.

Van Houten (1990) has observed further reductions by extending the back panel with a short tube running over the

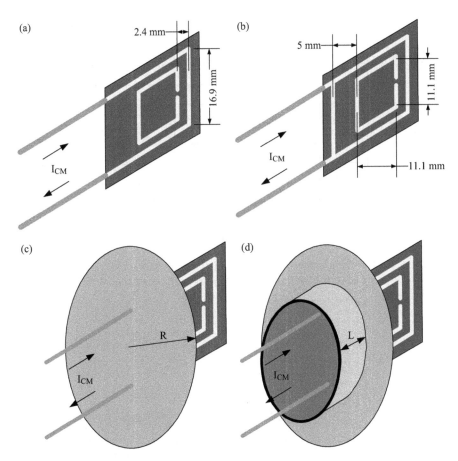

FIGURE 6.2 Current-carrying cable shields connected to a PCB: (a) cable shields connected via a long trace over the PCB; (b) cable shields connected with a short trace; (c) cables shields connected to a (circular) back panel; (d) back plane with tube around the cables.

cables (see Figure 6.2d). Figure 6.3 shows that the disturbance voltage is reduced when the dimension of the back panel increases. A similar effect is obtained by increasing the tube length. Obviously, the effect of the tube is the largest on a small back panel. In installation practice, application of a tube is particular useful for cable entries in buildings, because typically there is no room for a large backplane (Figure 6.4).

The results obtained from the analysis of circular back panels can be extended to rectangular plates. The coupling will depend on the orientation of the cable with respect to the aspect ratio of the plate, but the difference is limited to roughly a factor of 2 as shown in Table 6.1.

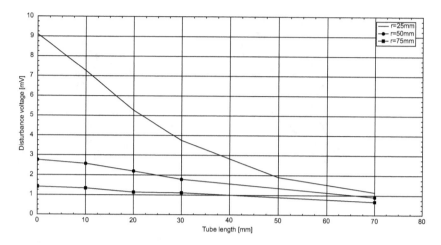

FIGURE 6.3   The disturbance voltage induced in a PCB caused by CM currents over a cable shield (see Figure 6.2c and d) decreases with increasing panel radius (R) and tube length (L).

FIGURE 6.4   Cable entry in a building via metal tubes reduces coupling of CM currents with equipment in the building.

6.1.3 Cabinets and electronic boxes

When the tube is not mounted on the back site of the panel, but on the front, an *enclosure* for the PCB is formed. Simulations show that for the round plane with a radius of 25 mm and a tube length of 10 mm the voltage drops with approximately a factor of 100 to 5.2 mV. This is a slight improvement compared to the tube around the wire (see Figure 6.3). Further improvements can be obtained by partial or full closing of the front (see Figure 6.5).

Table 6.1 Measured disturbance voltage when applying a rectangular back plane

Dimensions (mm)	Wire orientation	Disturbance voltage (mV)
Reference (Figure 6.2a)		610
400 × 50	Distributed over the length	8.2
	Distributed over the width	4.0
500 × 100	Distributed over the length	2.0
	Distributed over the width	1.2

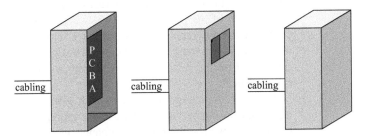

FIGURE 6.5 Extension of the back panel with side walls and a front cover further reduces the induced disturbance voltage.

Care has to be taken in the construction of cabinets and enclosures. Slits and holes increase the unwanted coupling. In particular, assemblies with coated panels are suspicious. Also doors on standard cabinets and lids on enclosure often leave gaps. For example, a door may have a rubber sealing which prevents any galvanic contact. The mandatory *grounding strap* for human safety is highly insufficient for EMC, electrically there remains a very large slit.

Also the orientation of the PCB with respect to the connection of the cables is important. In applications, the following situations are relevant, where the first is the worst and the latter is the best:

- PCB and connecting cables are assembled in the same orientation.
- PCB and connecting cables are perpendicular to each other.
  - PCB is mounted at some distance from the back panel.
  - PCB is mounted flat against the back panel.

In large enclosures, the coupling between the external and internal cabling will be more relevant than the direct coupling with the contained electronics. Therefore, the internal cabling must be carefully routed and kept close to grounded metallic structures. In this case, grounding does not mean a grounding strap, but a proper galvanic connection.

Analogous to the cable trays the best protection is found close to the back panel and in particular in the corners, assuming that the panels are connected over their full length. This was demonstrated by improving cable routing in a measurement cabinet. Initially a disturbance (DM) voltage of 800 mV was detected. By rerouting the cabling close to already present metallic structures and by connecting internal cables shields to these structures the voltage dropped by a factor of 10. When the door, which made electrical contact over its complete circumference, closed 10 mV was measured. Improvements in the cabinet design led to a reduction to 1.6 mV, a factor of 500 difference with the initial measurement.

Proper functioning of cabinets with open doors is very important during service activities. Therefore, only in exceptional cases one should rely on completely closed enclosures.

6.2 Connections to enclosures

6.2.1 Grounding rail

In Section 6.1.1, the benefit of a short grounding rail has been discussed. Such a rail can be used in combination with an enclosure as has been shown in Figure 6.1. For conductive enclosures, it is better to place the grounding rail on the outside of the cabinet as shown in Figure 6.6a and b, such that the large CM current circulates outside the cabinet.

The grounding rail must at least have a solid galvanic connection to the cabinet at both ends.

In configuration a, the grounding rail is mounted on top of, under, or at the sides of the cabinet. Coupling between the CM current and the DM voltage will remain due to leakage via the front opening. This can only be prevented when an expensive door is used with contact points all around its circumference. Another leakage path is provided by the hole through which the cable enters.

Often it is required that the outsides of the cabinet remain flush. In such a situation, a recess can be designed in which the grounding strip is mounted. Alternatively, a cabinet with multiple compartments can be used as shown in Figure 6.7. One compartment then is dedicated to connecting all cable shields and ground conductors to the grounding rail.

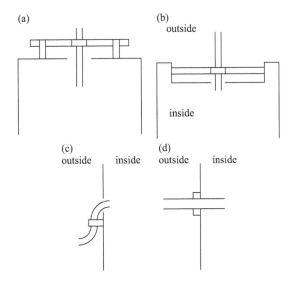

FIGURE 6.6   Connecting cable shields to an enclosure: (a) ground rail on top; (b) ground rail on top, recessed; (c) ground rail on back with full connectivity over its length; (d) feedthrough.

FIGURE 6.7   In a compartmental cabinet one compartment should be reserved for all grounding connections.

The best placement of the grounding strip is behind the cabinet, because it maximized the distance to the front opening and minimizes the coupling (see also Section 6.1.3). Figure 6.7c shows an alternative implementation grounding strip with direct attachment to the cabinet. The length of the cabling beyond the grounding strip should be minimized. Otherwise, the transformer effect will induce CM current on the cables penetrating the enclosure. Instead of bending the cables, straight-angled connectors can be mounted in the cabinet wall.

6.2.2 **Shielded connectors**

When mounting connectors in the enclosure wall is a viable option, the use of dedicated shielded versions should be considered. An example is given in Figure 6.8. The EMC performance depends on their quality and can best be judged by their *transfer impedance* (Section 3.4.5). If the transfer impedance is not specified, expert judgment is required: the more individual grounding contacts and the better they are distributed along the circumference, the better performance can be expected.

6.2.3 **EMC glands**

If disconnection of cables is not necessary or even unwanted (e.g., due to cost or performance limitations), special EMC glands (see Figure 6.9) can be applied. In general, these cable glands offer a very high EMC quality, because the cable shield makes contact with the cabinet all around its circumference. Quednau (2014) compared regular glands and EMC glands in an enclosure. Figure 6.10a shows the setup schematically. Cables 1 and 2 are fed through an EMC gland and cables 3 and 4 are fed through a regular gland. The shields of both pairs are connected to ground rail within the enclosure. When a 100 MHz CM current of 1 A is injected over cable 3, this leads

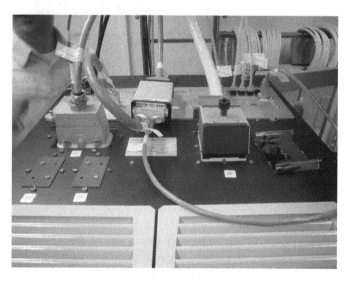

FIGURE 6.8 Cable shields are directly connected to the cabinet wall via shielded connectors.

FIGURE 6.9 For fixed installation special EMC feedthroughs or EMC cable glands provide optimal performance (Pflitsch).

to a current of 0.4 A over cable 4. When the same current is injected over cable 1, then only 20 mA flows over cable 2. The EMC glands give an improvement of a factor of 20. Results for other frequencies are shown in Figure 6.10b.

An objective measure for determining the EMC quality of cable glands is the *transfer impedance*. It describes the differential mode (DM) voltage caused by the CM current over the gland (compared with connectors in Section 3.4.1). An example, measured in the slightly modified setup of Figure 3.32, is presented in Figure 6.11.

6.2.4 EMC adaptors

If there is a need to replace a standard cable gland with an *EMC cable gland*, this can conventiently be achieved by the use of the EMC adapter as a lock nut or as an adapter (Figure 6.12a). With the use of a splittable EMC adapter one can connect a shielded cable at a later date without even having to disconnect any cabling (in case plastic glands are used). The cable must not be dismounted and withdrawn, because the two halves of the splittable EMC adapter will be positioned around the opened shield and joined together (Figure 6.12b).

6.2.5 Multi-cable transits

When many cable feed throughs are required it might be more cost efficient to accept slightly lower performance at improved speed of installation offered by *multi-cable transits* (MCTs). MCTs are flexible, conductive, elastomer blocks which can be stacked in a metalic frame. As shown in Figure 6.13, these blocks clamp around cable shields and connect them to each other and the enclosure. Blocks either contain contact springs, foil, or conductive paint (see Figure 6.14).

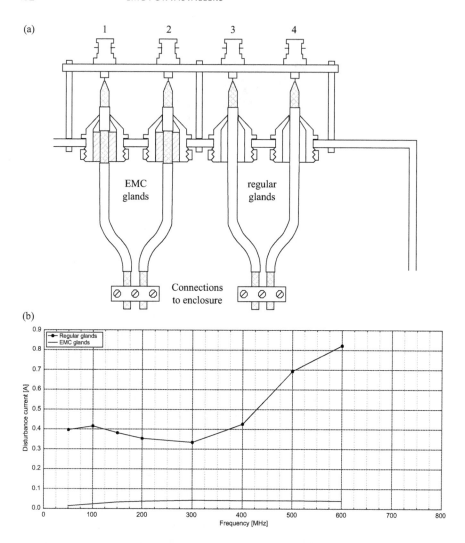

FIGURE 6.10 Coupling between cables without and with EMC glands (Quednau 2014): (a) schematic representation of test enclosure; (b) test results.

Figure 6.13 shows an ideal situation where the MCT frame is welded into a large metallic panel. MCTs are, however, also useful for a (central) cable entry in a building. In these cases, a connection is made to the construction steel used in a building. In these cases the performance will be less, because magnetic fields will leak through the openings such that the internal and external circuits will couple (Figure 6.15). Obviously, coupling can be reduced by the plate and tube measures described in Section 6.1.2.

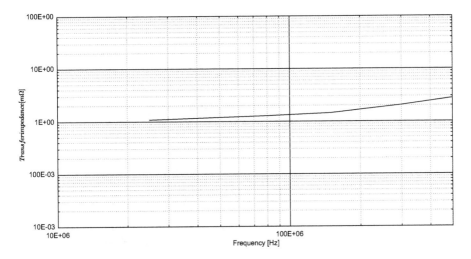

FIGURE 6.11   Transfer impedance of EMC gland (Pflitsch blueglobe TRI).

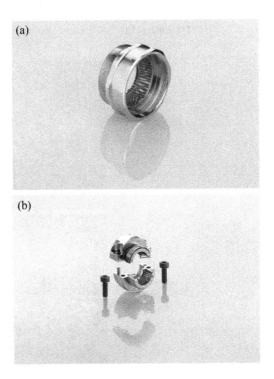

FIGURE 6.12 (a) EMC adapter for regular cable glands; (b) splittable EMC gland (Pflitsch).

FIGURE 6.13   MCT frame: (a) unmounted; (b) mounted.

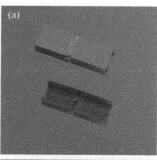

FIGURE 6.14   MCT block with (a) contact springs and (b) conductive paint and foil.

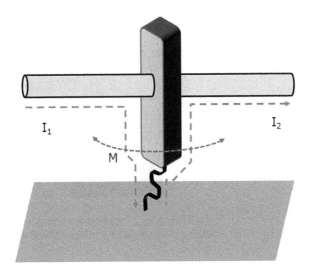

FIGURE 6.15   Magnetic field coupling for nonideal installation of an MCT frame.

6.2.6 Nonelectrical conductors

CM currents may not only flow over electrical cabling, but also over all other nonelectrical conductors. Examples are as follows:

- Copper water tubes
- Steel heating or sprinkler pipes
- Metallic coating on optical fibers
- Metal armor around flexible tubing (see Figure 6.16)
- Nonconductive tubing filled with conductive liquid (e.g., glycol solution in cooling of power electronic systems

Nonelectrical conductors have to be properly connected to the enclosure as well. In nearly all cases, the solutions for cabling described in Sections 6.2.3, 6.2.4 and 6.2.5 can be applied. In specific situations, custom-made solutions may be required.

6.2.7 Cable trays

Figure 4.19 proved that proper interconnection of cable trays is essential for their EMC performance. This is similarly true for interconnections between cable trays and enclosures. The best performance is given by U-shape interface. If this is not possible, a flat plate is second best, a third alternative is given by Mil-HDBK-419 (1987) where multiple parallel strips are used (see Figure 6.17).

In most countries, national safety standards mandate the independent grounding of cabinets with a dedicated grounding wire. In these situations, Merwe et al. (2011) showed that

FIGURE 6.16 Proper panel penetration of nonelectrical conductors: example of feedthrough of metal armored water cooling tubes.

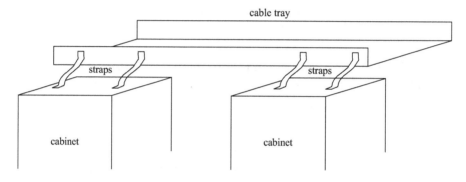

FIGURE 6.17 Connecting a cable tray to a cabinet with multiple parallel strips.

connection of this wire and the cable tray should be on the same site of the enclosure.

6.2.8 Corrosion   When two different metals are connected to each other a *potential difference* will arise between them. Any amount of conductive moisture (electrolyte) will lead to a chemical reaction, called *corrosion*. The residue created during corrosion may transform an initially good contact between both metals into a high impedance one over time. The rate of corrosion depends on the materials involved and the environment (see Tables 6.2 and 6.3).

If the difference in anodic index is too large for the targeted environment one can apply a surface treatment, such as plating. For example, one can apply chromium plating on aluminum if connection to copper has to be made in a normal environment.

Table 6.2 Maximum anodic index difference for various environments (Zygology)

Environment	Characteristic	Maximum difference in anodic index (V)
Outdoors	High humidity, salty	0.15
Normal	Covered area, no control of temperature and humidity	0.25
Controlled	Temperature and humidity regulated	0.50

Table 6.3 Anodic index of some relevant metals (Zygology)

Metal(s)	Anodic index (V)
Silver; high nickel–copper alloy	0.15
Nickel	0.30
Copper	0.35
Brass	0.40
Chromium-type corrosion-resistant steel	0.50–0.60
Chromium plated	0.60
Tin; tin-lead	0.65
Iron; low alloy steel	0.85
Aluminium (except 2000 series)	0.90–0.95
Galvanized steel	1.20
Zinc	1.25

In our experience, this worked well with hexavalent chromium, which due to RoHS now has been replaced by trivalent chromium. Trivalent chromium is however much less abrasion resistive and gets easily damaged during fastening leading to a direct contact between copper and aluminum, resulting in corrosion.

Similarly, one carefully has to check the surface treatment of connectors and filters. In a specific use case, a brass cabinet was replaced by an aluminum cabinet under the assumption that the feedthroughs were made from corrosion-resistant steel. The difference in anodic index of corrosion-resistant steel is 0.20 V with brass and between 0.25 and 0.45 V with aluminum. Both match the requirements for a controlled environment in which the cabinet was placed. Nevertheless, within a couple of months the shielding properties strongly deteriorated. After lengthy investigations, it was discovered that the filters where nickel plated. The difference in anodic index of nickel with brass is 0.10 V and with aluminum between 0.60 V and 0.65 V. The latter exceeds the requirements.

In case a proper surface treatment to either one or both metals is not possible, a potting solution could be used. Corrosion will not occur in an airtight, moisture-free environment.

6.2.9 Testing

Whatever method of feedthrough is chosen it is of utmost importance that the CM currents can flow between the feedthrough and the enclosure over the technical lifetime of the installation. Paint, being an insulator, has to be removed when installing an EMC gland or MCT frame. Alternatively, serrated washers can be used. If one cable is incorrectly installed it degrades the performance of the complete assembly. This can be tested with the setup depicted in Figure 6.18, which uses three current clamps. One clamp is used to inject current over the cable under test and the others are used to measure the common current before and after the enclosure wall.

The ratio between the measures of current probes 1 and 2 determines the EMC quality:

$$\text{Attenuation [dB]} = 20\log\left(\frac{I_2}{I_1}\right) \tag{6.1}$$

As example, we show a measurement on a multi-cable transit setup in Figure 6.19. The curve A shows the attenuation when

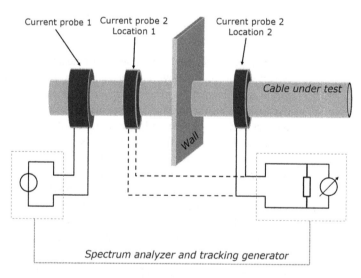

FIGURE 6.18 Test setup for determining the quality of cable shield connection to an enclosure.

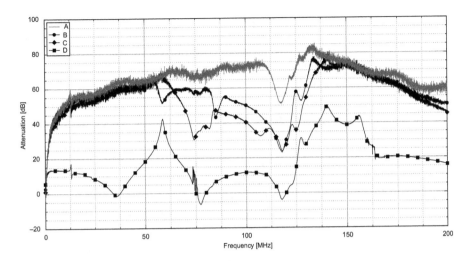

FIGURE 6.19   CM current rerouting with an MCT.

all cables are installed properly. The curves B and C show the attenuation when the connections are not optimal. When during installation it is forgotten to remove the outer insulation from the cable and on contact to the MCT is made at all, it will no longer function as intended: The curve D shows the strong degradation of the attenuation.

A practical approach to testing is to first measure a complete cable bundle (or sub-bundles), and if problems are found, measure each individual cable till the culprit is found.

6.3   Reducing CM currents

In certain situations, in particular during troubleshooting of an existing installation, it is hard to improve and/or replace existing connections to an enclosure. In those situations, additional grounding connections, shield wires, and/or EMC ferrites can be used to reduce current flow over sensitive cables and connections.

6.3.1   Ferrite

Ferrites are ferromagnetic materials, which have a similar relation between the *B*- and the *H*-field as steel: they are non-linear and have hysteresis. EMC ferrites are designed such that hysteresis is maximized, because this introduces losses, i.e., can be used to absorb disturbance energy. In order to reduce the complex behavior of ferrite *nonlinearity* and *saturation* from Figure 5.14 are ignored and only the *hysteresis* is considered as shown in Figure 6.20. In this case, the B-field follows the H-field with a material-specific and frequency-dependent phase delay δ (Getzlaff 2008). The electric equivalent diagram then can written as a resistor R and an inductor L in series with (Figure 6.21):

$$\tan \delta = R/2\pi fL \tag{6.2}$$

The larger δ the larger the loss and the larger R. Both R and L can be determined from supplier data sheets. Most often R and X $(=2\pi fL)$ are given as a function of frequency. Sometimes, suppliers separate material data and geometry factors by introducing the *complex permeability* μ and the inductance L_0, which is the self-inductance of the ferrite shape when μ_r would equal 1. The complex permeability μ is frequency dependent and typically graphs of the real μ' and imaginary part μ'' are given. An example is shown in Figure 6.22. R and L can be calculated as follows:

$$\mu = \mu' - j\mu'' \tag{6.3}$$

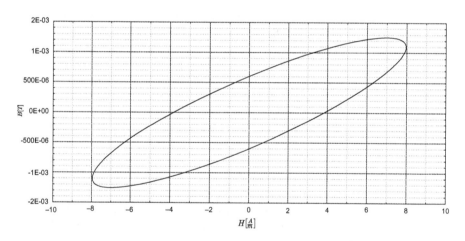

FIGURE 6.20 Approximated dynamic B–H curve with only hysterisis.

FIGURE 6.21 Standard series electric equivalent of a ferrite around a wire.

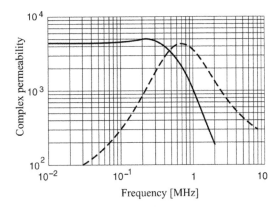

FIGURE 6.22 Real (μ': solid line) and imaginary parts (μ'': dashed line) of the complex permeability (μ) as a function of frequency (example).

$$R = 2\pi f L_0 \mu'' \tag{6.4}$$

$$L = 2\pi f L_0 \mu' \tag{6.5}$$

Some suppliers do not list L_0, but specify the inductance for a single turn, A_L. If the ferrite is completely closed and has no air gap as found in snap-on types, L_0 can be found as

$$L_0 = \frac{A_L}{\mu_i} \tag{6.6}$$

where μ_i is the *initial relative permeability* of the ferrite. In case the ferrite has an air gap, the *effective permeability* μ_e is specified and has to be used instead of μ_i:

$$L_0 = \frac{A_L}{\mu_e} \tag{6.7}$$

Yet another manner to specify *ferrites* (and *filters*) is *insertion loss* (IL). IL is determined by two measurements—one with the ferrite (or device under test [DUT]) and one without, as shown in Figure 6.23:

$$IL\ [\text{dB}] = 20\log_{10}\left(\left|\frac{V_{L,u}}{V_{L,a}}\right|\right) = 20\log_{10}\left(\left|\frac{Z_S + Z_L + Z_{DUT}}{Z_S + Z_L}\right|\right) \tag{6.8}$$

Typically, Z_S and Z_L are 50 Ω due to standard measurement equipment (see Section 3.4.3). Unfortunately, a 50 Ω source and load are not representative for most situations encountered in

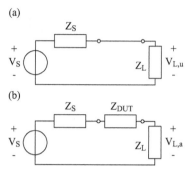

FIGURE 6.23 IL is determined by the ratio of the measured voltage at the load (a) without the DUT or unattenuated and (b) with the DUT or attenuated.

real installations, so actual realized IL can vary largely from the specified values. When $|Z_S|$ and $|Z_L|$ are known, real Z_{DUT} can be calculated,

$$|Z_{DUT}| = \left(10^{\frac{IL}{20}} - 1\right)|Z_S + Z_L| \tag{6.9}$$

Since only the absolute value of $|Z_{DUT}|$ can be determined it remains unknown if inductance or resistance is added. In case the inductive part of the ferrite is dominant, it will lower the resonance frequency and may even increase the problem instead of solving it.

Ferrites can increase the resistance (typically up to 300 Ω) of the CM or Intermediate Mode circuit, thereby decreasing the CM or intermediate (IM) current without impairing the transfer impedance of neither cable nor connection. Overall, this will result in a lower disturbance of the DM voltage. Ferrites have the largest effect in small loops, which have low impedance. Typically, it is more efficient to apply ferrites in the IM circuit than in the CM circuit. Ferrites can also be used on cabling within enclosures to further suppress disturbance voltages. In the demonstration described in Section 6.1.3 in the open cabinet the disturbance voltage could be lowered after rerouting the cabling by an additional factor of 1.3 by applying ferrites.

The impedance of a ferrite core may be further increased by increasing the number of turns (at the cost of installation convenience). Further more care has to be taken that the total current through the ferrite remains suffiently low. Close to the *saturation* point ferrites lose their effectiveness. In addition, care must be taken for operation at elevated temperatures. Above a certain threshold, the *Curie temperature*, ferrites lose their ferromagnetic property. Note that this is also the case when the temperature increase is caused by self-heating.

6.3.2 Resonant structures

In particular applications, typically involving narrow band RF transmitters and antenna installations, resonant structures can be galvanically of inductively connected to the cable shield to suppress CM currents on a specific radio frequency. Figure 6.24 shows a cable coiled as inductor with a capacitor connected to the shield at both ends, thus forming a band-block filter for CM currents, while not changing the DM behavior.

Figure 6.25a shows an alternative solution where a resonant structure is placed around the cable (Helvoort 2004). The resonant structure can be formed by a dielectric tube (circular or rectangular) with a conductive coating on both sites. On the one end the conductive coatings are shorted to each other, and on the other end, the structure is open. The CM current over the cable couples inductively with the structure around the cable which provides a high impedance at the resonance frequency. This concept is similar to sleeve or bazooka *baluns* used to connect unbalanced coaxial transmitter cables to balanced antennas (see Figure 6.25b; Lindenblad 1936). By choosing a higher dielectric permittivity for the resonant structure its length can be reduced (Peterson 1945), alternatively lumped capacitors can be soldered over the open joint.

6.3.3 Multipoint grounding

In Figure 6.26a, a sensor in a power electronics cabinet is connected to electronics in a control cabinet. To protect the measurement lead a grounding structure is used. In this example, the grounding structure has imperfect connections to the cabinets

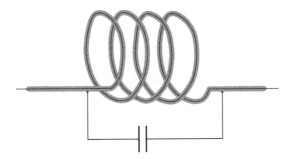

FIGURE 6.24 Cable coiled as an inductor with a capacitor connected to the shield at both ends. Near the resonance frequency this will provide a high impedance for CM currents.

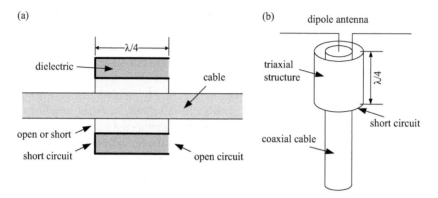

FIGURE 6.25 (a) Resonant structure inductively coupled to a cable forming a band-block filter; (b) sleeve or bazooka balun known from antenna technology.

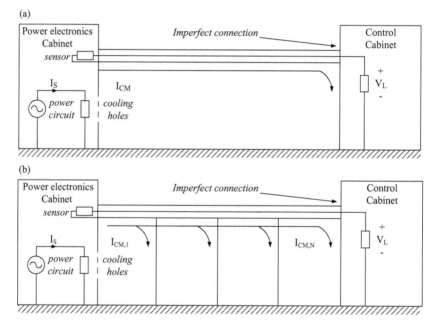

FIGURE 6.26 Grounding structure with imperfect connection: (a) due to leakage of the power cabinet, a CM current flows over to the grounding structure; (b) multipoint grounding provides additional return paths for the CM current and reduces the current flow over the imperfect connection.

and disturbance leaks from the power electronic cabinet causing a CM current. Via the transfer impedance of the grounding structure and the connection at the control cabinet a disturbance voltage is generated within the electronics cabinet. In case it is difficult to improve the grounding structure and its connection, *multipoint grounding* may be an option.

The current through the power circuit, I_S, induces a CM voltage, V_{CM}, in the loop formed by both enclosures and the grounding structure and the ground plane. This can be described by a local transfer impedance, $Z_{t,s}$:

$$V_{CM} = Z_{t,s}I_S \qquad (6.10)$$

When assuming that the grounding structure is long with respect to its height above the ground plane and that the first strap is sufficiently far away from the power cabinet, the voltage induced in the first loop of Figure 6.26b equals Eq. 6.10. Figure 6.27 gives a simplified representation for one additional strap and depicts the equivalent circuit diagram.

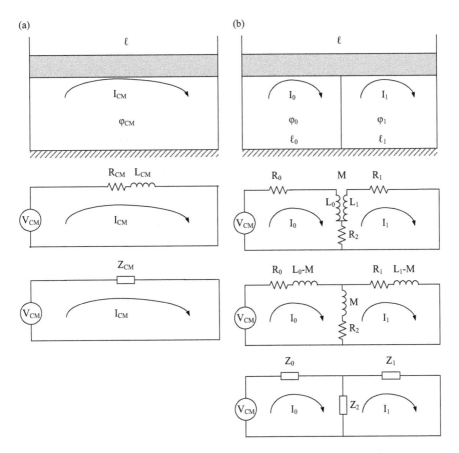

FIGURE 6.27 (a) Equivalent diagram of Figure 9.23a; (b) equivalent diagram of Figure 9.23b with one grounding strap.

The prerequisite for an effective grounding strap is $Z_2 \ll Z_0, Z_1$. In this case, we find a simple solution for I_1 as function of V_{CM} and I_S:

$$I_1 = \frac{Z_2}{Z_1} I_0 = \frac{Z_2}{Z_0 Z_1} V_{CM} = \frac{Z_2}{Z_0 Z_1} Z_{t,S} I_S \qquad (6.11)$$

In addition, we can express Z_0 and Z_1 as a function of x, where x is an arbitrary position of the ground strap along the grounding structures, as long as it is sufficiently far away from the cabinets:

$$Z_0 = \frac{x}{l} Z_{CM}$$
$$\qquad (6.12)$$
$$Z_1 = \left(1 - \frac{x}{l}\right) Z_{CM}$$

The current I_1 is minimal when Z_2 is minimal and $Z_0 Z_1 = \left(\frac{x}{l} - \frac{x^2}{l^2}\right) Z_{CM}$ is maximal, which both happen to be at $x = \frac{l}{2}$. In other words, the strap should be placed halfway the grounding structure (i.e., $l_0 = l_1$ in Figure 9.24b). Note that Z_2 is minimal when the mutual inductance M is minimal, since R_2 does not depend on the position. This mutual inductance can be approximated by Biot–Savart (Ramo et al. 1984).

In the above calculations we assumed a perfect cable or grounding structure. If we take a realistic cable with transfer impedance $Z_{T,cable}$ per meter, the total disturbance voltage in the control cabinet of Figure 9.23a is given by

$$V_{DM} = \left(l Z_{T,cable} + Z_{t,C}\right) I_{CM} \qquad (6.13)$$

When adding one additional strap, as in Figure 9.24b, we obtain

$$V_{DM} = l_0 Z_{T,cable} I_0 + l_1 Z_{T,cable} I_1 + Z_{t,C} I_1 \qquad (6.14)$$

Since $I_1 \ll I_0$ and $l_0 \approx l_1$,

$$V_{DM} \approx l_0 Z_{T,cable} I_0 + Z_{t,C} I_1 = l_0 Z_{T,cable} \frac{V_{CM}}{\frac{l_0}{l} Z_{CM}} + Z_{t,C} I_1$$

$$= l Z_{T,cable} I_{CM} + Z_{t,C} I_1 \qquad (6.15)$$

Comparing Eq. 6.13 with Eq. 6.15 shows that adding grounding straps only reduces the coupling effects due to the imperfect connection at the control cabinet. It does not alter the contribution of the cable transfer impedance. In case there is an imperfect connection at the power cabinet, the disturbance actually can be increased by adding the ground strap.

6.3.4 Shield wires

Cabling installed in buildings typically is not exposed to direct lightning strokes. This is different for cabling connecting buildings. Cabling which is not buried in the soil should be protected with proper parallel earthing conductors which have been discussed in Chapters 3–5.

Cabling buried in the soil is less prone to direct lightning strokes; however, a nearby lightning stroke will ionize part of the soil, which may lead to large currents over buried cables in particular when the soil is highly resistive (Sunde 1945). Chang (1980) has calculated the maximum radius over which the soil can be ionized by a lightning stroke as function of soil resistivity (see Figure 6.28):

$$100 \le \rho \; [\Omega m] \le 1100 \Rightarrow r_{max}(\rho) = \sqrt{\frac{1 \times 10^3}{2\pi e_0}} \cdot \sqrt{16\rho + \frac{2 \times 10^6}{\rho}}$$

$$(6.16)$$

For a conservative estimate of the radius, the breakdown voltage e_0 of (moist) soil is taken to be 1 MV/m and the maximum radius is found with $\rho = 100 \; \Omega m$: $r_{max} = 1.85\,m$. As a rule of

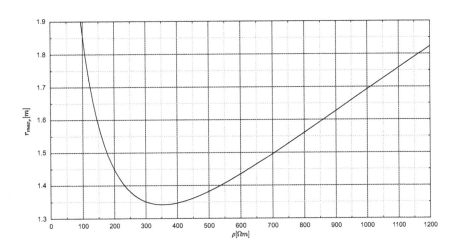

FIGURE 6.28 Maximum ionization radius of soil as function of the soil resistivity.

thumb, the length of a lighting discharge channel in soil is taken to be $2r_{max}$ (=3.7 m) without cable present and $3r_{max}$ (=5.5 m) with cable present. The latter is longer due to the electric field distortion caused by the conductive cable.

In case a cable is buried more than 5.5 m below the surface no additional protection is required. If it is less additional parallel earthing conductors are advised. Since tubes and trays are difficult, thus expensive, to install, application of a meshed ground grid or shield wires are typically better options. A *shield wire* is a noninsulated conductor which is placed next to or on top of the cable which requires protection. The recommended cross-section is 16 mm² (Chang 1980; Sunde 1968). Already with a single shield wire a strong reduction of lightning current flowing over the cable can be observed, as shown in Figure 6.29. Graph A of Figure 6.29 represents the absence of a shield wire, B uses 1 shield wire, C uses two and D uses three shield wires.

An example is given in Figure 6.30a where a cable is buried at a depth of 4.5 m. Around the cable a circle with $3r_{max}$ (=5.5 m) is plotted. This circle intersects with the surface in points A and B. Around these two points a circle is drawn with $2r_{max}$ (=3.7 m). When the shield wire is installed in the intersection of these two smaller circles the cable is assumed to be protected.

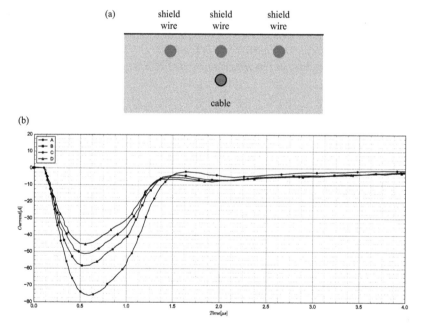

FIGURE 6.29 (a) Shield wires can protect buried cables against the effects of a lightning stroke; (b) effectiveness of shield wires. Based on (Yang, 2012).

Another example is given in Figure 6.30b; in this case the cable is buried at a depth of 2 m. The two circles, plotted around points A and B, no longer overlap and two shield wires are required. Note that the distance between these two shield wires is less than $2r_{max}$ (=3.7 m) as indicated by the circle plotted around point C.

6.3.4.1 *Stray currents*

Since shield wires are not insulated, they will corrode if DC currents from nearby traction systems or cathodic protection flow over them. Over time they will dissolve (see Table 6.4).

Electric currents do not go astray or wander around for they always return to their source. In the context of electrochemical corrosion, currents that flow through unintended or unforeseen metallic parts can pose a threat to metal parts. Examples are buried pipelines in the neighborhood of railway tracks. The railway tracks are part of the supply circuit and as such they are insulated from the earth in case of DC traction. As the

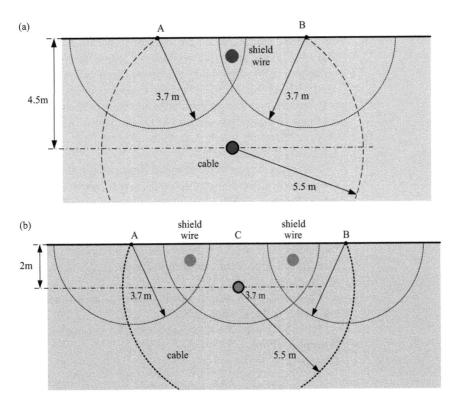

FIGURE 6.30 Examples of protecting a cable against lighting stroke with shield wires: (a) a cable buried at 4.5 m depth can be protected with a single shield wire; (b) a cable buried at 2.0 m depth requires at least two shield wires.

insulation is never perfect due to moist and the contamination of the ballast bed, some current will exit the rail into the earth. Buried conductors might prove to be an excellent path provided the now stray current, can enter and exit this parallel conductor. Note that the current direction will also change in railway operation, changing anodes into cathodes and vice versa.

The area where the current enters the buried (parallel) conductor is referred to as the "cathodic area" whereas the area where the current exits is referred to as the "anodic area". It is the anodic area where loss of material happens.

The above described the basic ingredients for stray current corrosion: two metal conductors that are separated by an electrolyte. The soil that separates the rail track from the buried conductor acts as the electrolyte.

The rate at which metals dissolve at the anode depends on many factors like the current: the stray current. In addition, factors that affect the electrolyte such as temperature, humidity, pH value and many more will also have their effect. This dissolution rate is derived from the definition of current as an not-changing electrical current that causes a deposition of 1.118 mg silver per second in a silver-nitrate solution (Teuchert 1959). This definition of the electrical current was used until August 14 1958 when the time-dependent definition was replaced by the electrodynamic force of $2 \cdot 10^{-7}\,\mathrm{mkg/s^2}$ between two parallel wires of infinite length, infinitesimal cross section and a $\mu_r = 1$ at a distance of 1 m, each meter cable length (Grafe 1967).

The general expression of metal dissolution rate, based on the pre-1958 definition of electrical current can be calculated using (Teuchert 1959):

$$G = a \cdot I \cdot t$$

In which G represents the deposit weight, a represents the electrochemical equivalent weight of the metal [kg/A · s] and t represents time [s].

This equation is still frequently used as an estimation of material loss due to stray currents.

Table 6.4 shows the dissolution rates for various materials (Philippow 1968)

Examples:

- The amount of deposited copper in 2 hours under a current of 7 A is: $(2 \cdot 7 \cdot 1.18576=)$ 16.6 g.

- Steel dissolves at a rate of $(1 \cdot 24 \cdot 365 \cdot 1.04176=)$ 9.13 kg per A each year.

Table 6.4 Stray current corrosion rate of different metals per 1 A per year

Element	Valence	A [g/A·h]	Material loss [kg/A·year]
Ag	1	4.02454	35.25
Al	3	0.33538	2.94
As	3	0.93152	8.16
As	5	0.55891	4.9
Au	1	7.35668	64.44
Au	3	2.45223	21.48
Ba	2	2.56216	22.44
Be	2	0.16825	1.47
Bi	3	2.59896	22.77
Bi	5	1.55938	13.66
Br	1	2.98132	26.12
Ca	2	0.74761	6.55
Cd	2	2.09677	18.37
Ce	3	1.74255	15.26
Cl	1	1.32275	11.59
Co	2	1.09931	9.63
Cr	3	0.64676	5.67
Cr	6	0.32338	2.83
Cs	1	4.9583	43.43
Cu	1	2.37152	20.77
Cu	2	1.18576	10.39
Fe	2	1.04176	9.13
Fe	3	0.69451	6.08
H	1	0.037605	0.33
Hg	1	7.4839	65.56
Hg	2	3.74195	32.78
J	1	4.73484	41.48
Ir	4	1.80095	15.78
K	1	1.4585	12.78
Li	1	0.2589	2.27
Mg	2	0.45364	3.97
Mn	2	1.02458	8.98
Mn	4	0.51229	4.49
Mo	6	0.59689	5.23
N	3	0.17419	1.53
N	5	0.10452	0.92
Na	1	0.85792	7.52
Ni	2	1.09474	9.59

(*Continued*)

Table 6.4 (*Continued*)   Stray current corrosion rate of different metals per 1 A per year

Element	Valence	A [g/A·h]	Material loss [kg/A·year]
O	2	0.29845	2.61
Os	4	1.78601	15.65
P	5	0.23115	2.02
Pb	2	3.86506	33.86
Pb	4	1.93258	16.93
Pd	4	0.99513	8.72
Pt	4	1.8208	15.95
Rb	1	3.18889	27.93
Re	7	0.99292	8.7
Rh	4	0.95978	8.41
Ru	4	0.9485	8.31
S	2	0.59801	5.24
S	4	0.29901	2.62
S	6	0.19934	1.75
Sb	3	1.51411	13.26
Sb	5	0.90847	7.96
Se	6	0.49094	4.3
Sn	2	2.21409	19.4
Sn	4	1.10705	9.7
Sr	2	1.63455	14.32
Te	6	0.79343	6.95
Ti	4	0.44674	3.91
Tl	3	2.54164	22.26
U	6	1.48023	12.97
V	5	0.38015	3.33
W	6	1.14404	10.02
Zn	2	1.21952	10.68

6.4 DM protection

Once disturbances have coupled into the DM circuit, only two options for protection are left: filters and overvoltage (surge) protection. Filters are the preferred option; however, they only can be applied if the frequency range of the disturbance differs from the intended signal range. If this is not the case, surge protection should be considered. Activation of the surge protector should however occur only incidentally like at the start-up of an electrical engine or resulting from a lightning strike. When activated, surge protection clamps the intended signal and the contained information is lost. Further it introduces fast and sharp voltage peaks. Modern equipment should be able to withstand these peaks, if not a low pass filter should be added.

6.4.1 Filters

We will consider the two most relevant types in EMC: mains filters and feedthrough filters.

6.4.1.1 Mains filters   Mains filters are low-pass filters which are used to block propagation of higher frequency disturbance over mains connections. Figure 6.31 shows a typical design. Connections to these filter include "line" (L), "neutral" (N) and "protective earth" (PE). The *IL* is defined as the power measured at the output of the filter without the filter divided by the power measured with the filter present it can be rewritten using the signal amplitude (see Figure 6.23 and Eq. 6.8). As with ferrites suppliers typically measure IL when both sites of the filter are terminated with 50 Ω (see Figure 6.32a and b; the filter from Figure 6.31 is depicted as block diagram). In this case, $V_{L,u} = V_0/2$ can be substituted in Eq. 6.8, where V_0 is the open-circuit voltage of the 50 Ω generator and *IL* can be written as follows (CISPR 17 2011):

$$IL\ [\text{dB}] = a_e\ [\text{dB}] = 20\log_{10}\left(\left|\frac{V_0}{2V_{L,a}}\right|\right)$$

$$= 20\log_{10}\left(\left|\frac{2Z_0 + Z_{DUT}}{2Z_0}\right|\right) \tag{6.17}$$

In this setup, the IL conveniently becomes identical to the attenuation a_e between the input and the output, which can expressed with S-parameters (see Section 3.4.3):

$$IL\ [\text{dB}] = A_e\ [\text{dB}] = -20\log_{10}\left(|s_{21}|\right) \tag{6.18}$$

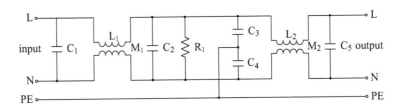

FIGURE 6.31   Typical mains filter design.

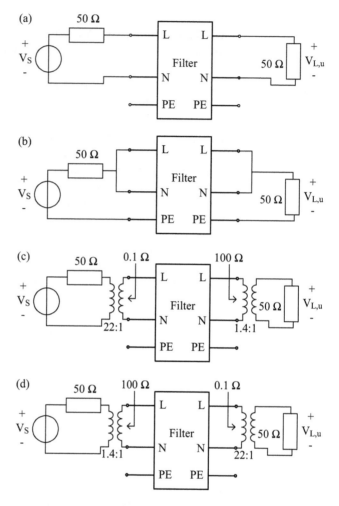

FIGURE 6.32 IL measurement setup. (a) DM with 50 Ω termination; (b) CM with 50 Ω termination; (c) AWCM, DM 0.1 Ω at source, 100 Ω at load; (d) AWCM, DM, 0.1 Ω at source, 100 Ω at load.

Since there are three connections two different attenuation curves can be defined:

- *DM* (or *normal mode* or *symmetrical mode*): measured between line and neutral
- *CM* (or *asymmetrical mode*): measured between the protective earth and short-circuited line and neutral

Figure 6.33 shows a typical example of both attenuation curves. Since neither the mains network nor the connected source or

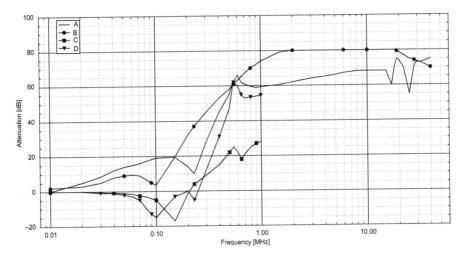

FIGURE 6.33 IL measurement. (a) DM with 50 Ω termination; (b) CM with 50 Ω termination; (c) AWCM, DM 0.1 Ω at source, 100 Ω at load; (d) AWCM, DM, 0.1 Ω at source, 100 Ω at load.

load, is likely to have a 50 Ω impedance, this attenuation measurement has limited value in predicting the performance of the filter in the actual circuit. This problem can be overcome with the "approximate worst-case method" (AWCM) which is provided as an alternative measurement in CISPR 17 (2011).

The AWCM measures the DM IL with 0.1 Ω termination on the input (also labeled as source) and 100 Ω at the output (also labeled as load). Then, the measurement is repeated with the impedances swapped (see Figure 6.32c and d). The DM measurements with AWCM (curves C and D) have been limited to 1 MHz, because at higher frequencies the CM current becomes dominant. The AWCM measurement results have much better predictive value for the actual in circuit behavior (Schaffner 1996).

In particular, the potential amplification of disturbances between 20 and 250 kHz in real applications, with peaks at 100 and 150 kHz, cannot be derived from the measurements with 50 Ω terminations.

As discussed in Section 6.2 connections to enclosures are only effective if good circumferential contacts are ensured. Painting or isolating gaskets have to be prevented. This is once more shown in Figure 6.34 which depicts the CM IL of the same mains filter under different grounding conditions:

- Curve A: proper circumferential connection of the filter to the enclosure

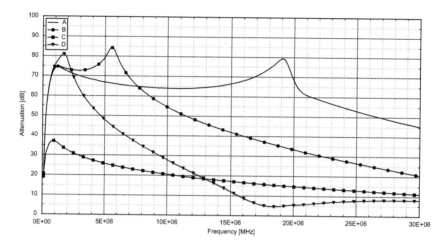

FIGURE 6.34   IL of a filter depends on the quality of its connection to the enclosure: (a) proper connection; (b) 16-cm-long pig tail; (c) no connection; (d) 1-m-long pig tail.

- Curve B: connection of the filter housing via a 16-cm-long pig tail made of 1 mm² wire
- Curve C: no galvanic connection between filter and enclosure
- Curve D: similar to situation B, but with a pig tail of 1 m long

6.4.1.2 Feedthrough filters   For connecting *unbalanced* signal wires the best filter is a feedthrough filter. This filter has a coaxial structure with a capacitance flush to enclosure wall, acting as the feedthrough and an inductance in series as shown in the schematic diagram of Figure 6.35. Due to their construction feedthrough filters' proper mounting ensures minimal coupling between both ends of the filter.

6.4.1.3 Nonideal components   Common practices in electronics are to create designs based only on the characteristics of ideal components: resistors, inductors, and capacitors are independent of frequency and voltage. For designing or selecting a filter this approach is not appropriate and the design has to be based on real, *nonideal component*, behavior. As example the capacitor C_2 and R_1 in the mains filter of Figure 6.31 for a loop which might couple with coil L_1 or L_2. The dielectric C_2 typically will be frequency dependent, and if the resistor R_1 is a wire-wound type, it will also act as an inductor at higher frequencies. The windings of transformers L_1 and L_2 may shorted by parasitic capacitances formed

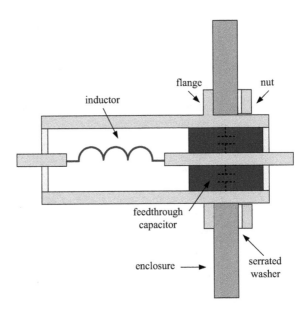

FIGURE 6.35   Schematic representation of a feedthrough filter.

by the windings themselves. In filters often toroidal coils are used to reduce stray magnetic field, thus minimizing unwanted coupling. Figure 6.36 shows two winding configurations for these transformers: *sectional* and *bifilar*. With sectional wiring the stray field is less than 1%, but with the bifilar winding even less; however, at the cost of increased capacitive coupling between both coils thus lowering the bandwidth. Further, it is important that current remain sufficiently low to prevent *saturation*, which reduces the inductance and the overall attenuation of the filter.

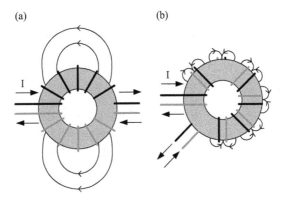

FIGURE 6.36   Toroidal ferrites reduce stray magnetic fields: (a) sectional windings; (b) bifilar winding.

Table 6.5   Characteristics of overvoltage protection components

Type	Spark gap gas discharge tube	MOVs	Transient voltage suppressor diode
Allowable current/energy dissipation	High	Medium	Low
Reaction speed	Slow (μs)	Average (ns)	Fast (ps)
Leakage current	None	Small, increases over lifetime	Small
Capacity	pF	nF	pF
Maximum signal frequency	High	Medium	High
Lifetime	Very long	Limited, degrades over lifetime	Long
AC or DC	AC and DC, may not extinguish with DC	AC and DC	DC, AC with anti-series pair

6.4.2 Overvoltage protection

Theoretically, overvoltage protection, also known as surge protection, limits the maximum voltage in the differential circuit to a predefined value. In reality systems with limiters show much more complex behavior and should always be treated with caution. A variety of *surge protection devices* (SPDs) are nowadays on the market, such as *spark gaps*, *gas discharge tubes* (GDTs), *metal oxide varistors* (MOVs), and *transient voltage suppression* (TVS) diode. Key characteristics are given in Table 6.5 and a short description is given in the sections that follow.

6.4.2.1 Spark gaps and GDTs   A spark gap consists of two electrodes at such a distance that a flash over will occur when the voltage exceeds the specified limit. The discharge ionizes the air causing a low impedance path through which current flows. A certain minimum holding current is required to sustain the plasma. If the voltage over the spark gap has become sufficiently low the current drops below this minimum and the gap extinguishes itself. However, if a DC power source is shortened by the spark gap and this source is sufficiently strong, the gap may never extinguish. Open versions, where the plasma can escape to the sides, may be more appropriate for these applications than their closed counterparts. Some countries allow placement of a spark gap before the metering installation.

A gas discharge tube is a closed variant of a spark gap filled with a specific gas to change the discharge voltage. Neon lamps are a common example, often used in low voltage applications.

6.4.2.2 Varistors A varistor (i.e., varying resistor) acts as a voltage-dependent resistor which has a high resistance at low voltages. The resistance increases when the voltage is increased. The most common-type varistors are MOVs, which are typically based on sintered granular zinc oxide. The contact points between the grains act as diode junctions. Due to the random orientation of the grains, they are suitable for both AC and DC applications. A varistors will not short the circuit, but clamp the voltage at a certain threshold. In a power line this may cause the varistor to heat up which further lowers its resistance and finally may lead to a failure. This process is called *thermal runaway.* Such failure may be prevented by adding a *thermal fuse* in series. Over time MOVs degrade beyond acceptable limits and have to be replaced.

6.4.2.3 TVS diode A TVS diode is a Zener diode specifically designed for overvoltage protection. Clamping values range between 3 and 200 V. The breakdown voltage of a Zener diode decreases with temperature. The maximum current rating is the maximum current at the rated voltage the diode can handle without breaking down. The power rating equals the product of maximum current and rated voltage. TVS is often used to protect sensitive integrated circuits against *electrostatic discharges* and even may be built in.

6.4.2.4 Coordinated surge protection The various types of overvoltage protectors each have their advantages and disadvantages. In *coordinated surge protection,* various types are put in parallel to optimize the overall protection of the load. As an example, a varistor can be combined with a spark gap. When a transient overvoltage arrives at this combination within nanoseconds the varistor will respond and clamp the voltage, then within microseconds the spark gap will discharge. This will lower the voltage over the varistor which then will switch off, thus preventing thermal runaway. Optimized varistor and spark gap combinations are commercially available. Also alternatives which combine different types of MOVs are offered commercially; as discussed for individual varistors, they do not solve the problem of thermal runaway.

*6.4.2.5 **Lightning protection zone concept*** *Lightning protection zones* (LPZs) are based on a risk assessment of the impact and likelihood of a *lightning* strike in the specific zone (IEC 62305-4, 2012). Three zones are distinguished (Figure 6.37):

- LPZ 0: subject to full or partial lightning current. No attenuation of electromagnetic fields caused by lighting. Typically, this is an outdoor situation.

- LPZ 1: partial lightning current, limited by current distribution, isolating interfaces, filters, or SPDs. Some attenuation of electromagnetic fields by unintended shielding. Typically, this is an indoor situation where shielding is provided by the steel used in the building.

- LPZ 2: residual lightning current, limited by current distribution, isolating interfaces, filters, or SPDs. Typically, this is a metallic enclosure.

Risk management is used to determine the zone classification (IEC 62305-2 2012).

When surge protectors are used in the lighting protection zone this is a variation on *coordinated surge protection*, because in protectors in each zone must match with the characteristics of the devices in the other zone.

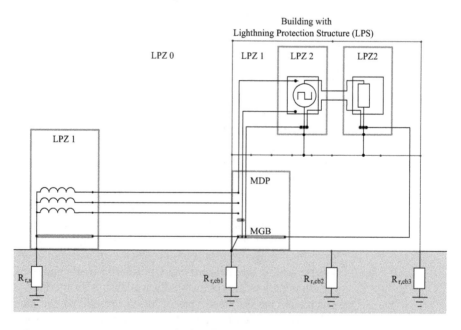

FIGURE 6.37 LPZ concept applied to the reference installation in Figure 2.24.

6.4.2.6 Limitations of overvoltage protection In this chapter, we mainly described passive measures to improve the EMC performance of installations. When properly installed these solutions are very robust and require little maintenance. Overvoltage protection is an essential part of proper EMC design in a complex system or large installation. However, their use should be minimized, because they increase the need for monitoring and maintenance leading to increased cost over the total lifetime of the installation or system.

Furthermore, as with passive measures, great care has to be taken when installing these components. When comparing Figure 6.38a with Figure 2.14, it will be obvious that a current through the surge protection device will induce a voltage in the output circuit due to resistive and capacitive coupling. Dedicated wiring between the input and output terminals will reduce the resistive coupling (Figure 6.38b). The magnetic coupling can be reduced by carefully routing the connection wires (Figure 6.38c).

Another way to reduce the effects of this coupling is to add a properly mounted filter behind the SPDs. When a surge protective devices sets on, most often it creates sharp spike with high frequency content. The filter will protect the load against this spike.

In regular installations overvoltage protection in the mains supply lines is not essential. International Electrotechnical Commission (IEC) has identified common surges on power lines (IEC 61000-4-5 2014) and demands via product standards that connected equipment can withstand them (e.g., IEC 11160 of information technology equipment or IEC 60601 for medical elecrical equipment). For system designers this is of course different; they are the ones making sure that IEC demands are met and may opt for SPDs.

In general, interconnecting grounding cables and structures via SPDs, or in other words using SPDs in CM circuits, is

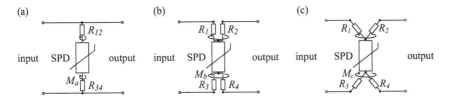

FIGURE 6.38 Methods for connection a surge protection device. (a) two-point connection to SPD (common leads between the input and the output); (b) four-point connection; (c) four-point connection with slanted terminals. The latter has the least coupling between the input and the output.

conceptually strange: Just at the critical moment the connection between both circuits is made anyway. It is cheaper and
requires less maintenance to make a permanent connection.
There are, however, exceptions to this rule; one of them has
been discussed in Section 3.2.2.

CHAPTER SEVEN

Barriers against radiated disturbances

Throughout the book, the common bonding network is propagated as the best overall solution for complex installations and large systems. The advantage of this network is that it keeps voltages induced in critical cables low by allowing currents to flow. In the previous chapter, it was shown how enclosures can be used to reroute these currents and keep them away from more susceptible zones and sensitive electronics. This chapter continues with *barriers* against radiated disturbances. At low frequencies, these disturbances are independent electric or magnetic fields. Therefore, they are the primary topic of this chapter.

7.1 Reducing magnetic coupling

Time-varying magnetic fields induce voltages in nearby loops as already shown by Figure 2.14. In most situations, these fields are generated unintentionally by current carrying cables, in other applications they are stray fields leaking from, e.g. transformers. Preventing 50 or 60 Hz hum caused by the transformer of a linear power supply was a common design challenge in audio amplifiers before the mass introduction of switched mode power supplies (Robbins 2017).

Time-varying magnetic fields caused by currents are not limited to AC (alternating current) installations. They are encountered as often in DC (direct current) installations. It is a regular misconception that DC stands for 0 Hz, which is not the case: DC means that the charges move in one direction only, not that they move in a constant pace. In other words, DC currents vary in time; thus, they generate time-varying magnetic fields and will cause induction (van Helvoort 2017).

Also moving or rotating a conductive loop through a static magnetic field will cause induction. This effect may be encountered near inductive ovens or magnetic resonance imaging (MRI) scanners.

There are four possibilities to reduce the *magnetic coupling* between two circuits:

- Change the orientation of the source with respect to the victim.
- Increase the distance between the source and the victim.
- Influence the shape and amplitude of the stray field.
- Apply shielding.

Combinations of the mentioned options are also possible.

7.1.1 **Orientation**

Magnetic fields only induce a voltage in a loop, when the field lines penetrate the loop of the field. When the loop is placed flush with the direction of the field, neither penetration nor induction will occur. When we place a current carrying loop orthogonal to another loop, no voltage will be induced in the second loop, as shown in Figure 7.1.

Figure 7.2 shows a temperature sensor in an inductive oven. It has been mounted such that no field lines penetrate the area formed by both measurement leads and the conductive object under test. During operation of the oven the temperature can be measured error-free. Also, in the earlier mentioned example about audio amplifiers, it was often sufficient to rotate the transformer to remove the hum.

7.1.2 **Distance**

Any magnetic field is strongest near its source and rapidly decreases when the distance increases. Initially, in the example of the inductive oven from Figure 7.2, the electronic conversion unit, labeled as "Detector B" in Figure 7.2b, was placed close to the oven. During operation, it registered "impossible" temperatures, while the temperature measured using the more compact unit, labeled as "Detector A," showed the expected values. This problem was solved by relocating the unit further away from the oven.

7.1.3 **Stray field reduction**

In the design of complex installations and large systems, we often are in charge of both the source and the victim. Design choices, i.e., potential mitigation or avoidance measures in the source–victim matrix (see Table 1.1), should be optimized

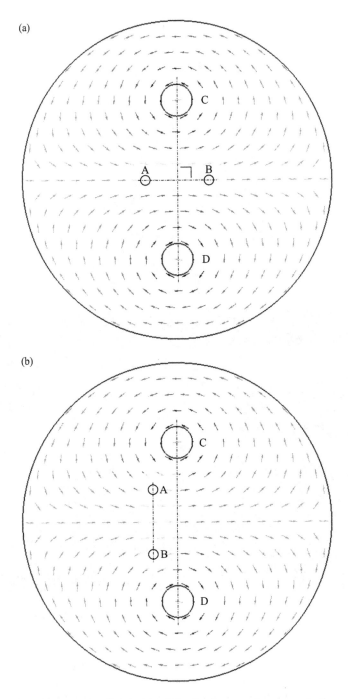

FIGURE 7.1 Magnetic coupling is (a) minimal when two loops are in the same plane and (b) maximal when two loops are perpendicular to each other.

(a)

(b)

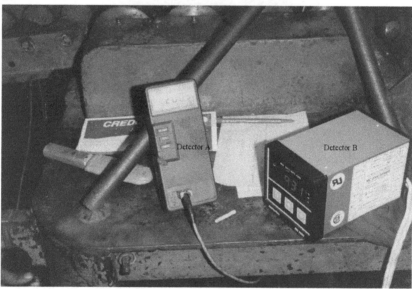

FIGURE 7.2   Inductive oven with (a) a temperature sensor and (b) two electronic detectors with difference in susceptibility.

for the integral solution. This may involve measures at the source, the victim or both. As an example yoke-type transformers (the source) are cheaper than toroidal transformers. This price difference, however, may be offset by the required shielding measured at the victim, because toroidal transformers produce less stray field than yoke types (compare Figure 6.36a with Figure 7.3). Typically the stray field of toroidal coils is less than 1% and can even be further reduced by applying bifilar windings (Figure 6.36b). In filters, toroidal coils can reduce coupling between input and output, thereby increasing the barrier between two zones.

Also cables can be routed such that stray fields are reduced. Figure 7.4 shows three single-phase power cables. The triangular arrangement of the cables leads to a 26% reduction stray field compared to the flat configuration (ABB 2010). A further

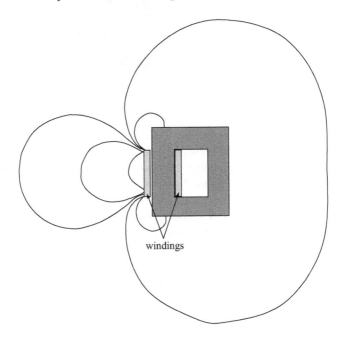

windings

FIGURE 7.3 Stray fields around a yoke type transformer.

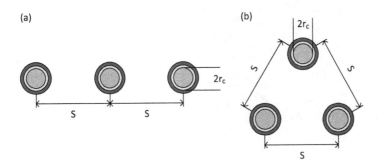

FIGURE 7.4 Three single core cables: (a) flat configuration; (b) triangular configuration.

FIGURE 7.5   Multi-core cable (in this case three phases with neutral) minimizes stray fields.

reduction could be obtained by selecting a three-phase cable (Figure 7.5), because the loops between the conductors are then smaller. Twisting wires also can reduce coupling, as shown in Figure 2.19.

A cable with four conductors (three phases and neutral) and a screen as protective earth (PE) will have in practical situations a lower stray field than a cable with five cores (three phases, neutral, and PE). In the latter case, the currents through the phase and neutral conductors couple more easily with loop formed by the PE conductor and nearby construction steel, as shown in Figure 7.6. This will give rise to a common mode current over the cable. This effect is only important if very

(a)   (b)

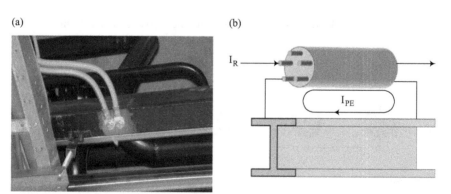

FIGURE 7.6   The PE conductor forms a loop with construction steel: (a) sketch; (b) image. Symmetry of the cable is important.

sensitive equipment, such as electron microscopes, MRI scanners, or audio recording equipment (Waldron 2007) have to be installed.

7.1.4 **Shielding**

If the previous measures are insufficient, shielding is the last resort. In general, shielding is both costly and difficult to maintain over the lifetime of installations and other important electronic equipment. Passive shielding may be altered unintentionally such that it loses its function and active shielding is not often practical.

7.2 **Passive shielding**

Enclosures can be used to attenuate disturbing magnetic and electric fields on the inside in order to shield the environment from a disturbance source. They can also be used to attenuate disturbing fields in the environment outside the enclosure to protect sensitive electronics.

In Section 3.4.1, the *shielding effectiveness* (SE) of cables was defined. A similar definition can be used for enclosures. The shielding effectiveness is the ratio of field strength at a certain location, measured without and with the shield present.

$$SE_H = 20\log_{10}\left(\frac{H_{\text{without shield}}}{H_{\text{with shield}}}\right)$$

$$SE_E = 20\log_{10}\left(\frac{E_{\text{without shield}}}{E_{\text{with shield}}}\right)$$

(7.1)

The higher the value obtained for *SE*, the higher the attenuation of the fields and the better the shielding performance. Figure 7.7 shows a generic shielding effectiveness curve for the electric and for the magnetic field. The most difficult is shielding magnetic fields at low frequencies.

The attenuation of realistic enclosures can only be determined via computer simulations or measurements. Nevertheless, good predictions can be made by approximating an enclosure, by two parallel plates, a cylinder or a sphere. Two different methods will be described:

- Predictions based on analytic approximations of the full field equations (Section 7.2.1)
- Predictions based on transmission line approximations of the full field equations (Section 7.2.2)

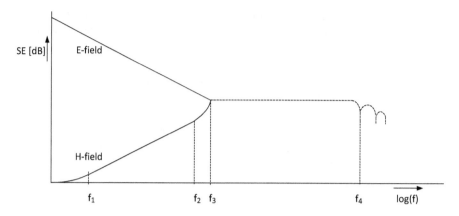

FIGURE 7.7 Generic shielding effectiveness curve for electric and magnetic fields. Frequencies f_1, f_2, f_3, and f_4 are given in Table 7.1.

Table 7.1 Magnetic shielding effectiveness for an enclosure approximated by a sphere

Description	Frequency range	SE_H SHE (dB)
Magnetostatic	$f < f_1 = \dfrac{1}{\pi \mu_0 \sigma r d}$	$20 \log\left(1 + \dfrac{2\mu_r d}{3r}\right)$
Induced eddy currents	$f_1 < f < f_2 = \dfrac{1}{\pi \mu_0 \mu_r \sigma d^2}$	$20 \log\left(1 + \dfrac{2\pi f \mu_o \sigma r d}{3}\right)$
Skin effect	$f_2 < f < f_s$	$20 \log\left(\dfrac{r}{3\sqrt{2}\mu_r \delta} e^{d/\delta}\right)$
Slits, slots, and holes	$f > f_s$	Section 7.2.6
Resonances	$f > f_4 = \dfrac{3 \times 10^8}{2h}$	Section 7.2.7

7.2.1 **Analytic approximations**

A cubically shaped enclosure of a given volume can be approximated by a sphere with identical volume. The following expression then can be used for estimating the shielding effectiveness over a large frequency range, until the wavelength becomes smaller than the wavelength of the field (King 1933):

$$SE_H = 20 \log\left(\cosh(kd) + \frac{1}{3}\left(\frac{kr}{\mu_r} + \frac{2\mu_r}{kr}\right)\sinh(kd)\right) \quad (7.2)$$

$$k = \frac{1+j}{\delta}, \text{ with } \delta = \sqrt{\frac{2}{\omega \mu_0 \mu_r \sigma}} = \sqrt{\frac{1}{\pi f \mu_0 \mu_r \sigma}} \quad (7.3)$$

In this equation, r is the radius of the equivalent sphere and d is the wall thickness of the enclosure. The frequency is given by f and the radial frequency by ω. The material properties are the conductivity σ and the permeability. μ_0 is the permeability of vacuum and μ_r the relative permeability. Figure 7.8 shows the calculated magnetic shielding effectiveness for two aluminum, one steel, and one stainless steel enclosure. The dimensions are $11.5 \times 9.2 \times 5.1$ cm, which is based on an experiment which will be described later (see Figure 7.21). The wall thickness was chosen 1 mm and 2 mm respectively. The box was completely closed (no slits or other gaps).

For design purposes often a specified shielding effectiveness has to be met and a suitable material of sufficient thickness has to be found. In this case, it is useful to approximate Eq. 7.2 by simpler expressions in each frequency area. These expressions are summarized in Table 7.1 and discussed in the sections that follow.

7.2.1.1 *Magnetostatic shielding*

Magnetostatic shielding is required when fields with a very low frequency, including 0 Hz, have to be attenuated. For these frequencies, $\cosh(kd) \approx 1$ and $\sinh(kd) \approx kd$ and $2\mu_r/k_r \gg k_r/\mu_r$, so Eq. 7.2 can be reduced to

$$f < f_1 = \frac{1}{\pi\mu_0\sigma rd} \Rightarrow SE_H = 20\log\left(1 + \frac{2\mu_r d}{3r}\right) \qquad (7.4)$$

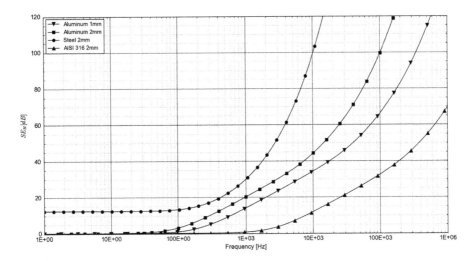

FIGURE 7.8 Calculated magnetic shielding effectiveness for a box of $11.5 \times 9.2 \times 5.1$ cm.

As can be observed, magnetostatic shielding is only obtained when ferromagnetic materials are applied as enclosure. These kind of materials have the capability to "conduct" (or "shunt") the magnetic field thereby reducing the field strength within the enclosure. This is graphically depicted in Figure 7.9.

Equation 7.2 shows that the effective magnetostatic shielding poses the following requirements on the enclosure dimensions and the relative permeability of the shield:

$$r \ll \mu_r d \tag{7.5}$$

If the cabinet becomes too large with respect to the shield thickness, fields may leak out of the shields, thereby lowering the shielding effectiveness. For regular plate steel commonly used in enclosures and equipment cabinets, the relative permeability will be approximately 120 in low fields. However, this value actually depends on the strength of the magnetic field and may increase up to 2500 for larger fields, i.e., the observed relative permeability depends on the field strength.

A steel box with dimensions $11.5 \times 9.2 \times 5.1$ cm and a wall thickness of 2 mm has a shielding effectiveness of 12.5 dB, which means that 50 Hz fields are reduced with a factor of 4. If a larger reduction factor is needed we can increase the wall thickness, or use mu-metal. Mu-metal is nickel-iron alloy with a very high relative permeability up to 50,000. In contrast to steel, the permeability rapidly drops with increasing field strength and, at increasing frequency, even at 50/60 Hz (Luker 1998).

Figure 7.10 shows an application of magnetostatic shielding to reduce the stray field of a large superconductive magnet. It is noteworthy that the shielding has not to be closed completely around the circumference.

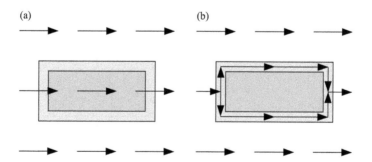

FIGURE 7.9 Enclosure in a static magnetic field: (a) nonmagnetic walls; (b) ferromagnetic walls.

FIGURE 7.10   Magnetostatic shielding of an MRI system in Leiden in 1986.

7.2.1.2 Shielding due to induced eddy currents   When the frequency becomes higher than the magnetostatic limit, such that $k_r/\mu_r \gg 2\mu_r/k_r$, eddy currents start to flow over the enclosure leading to an increase in magnetic shielding effectiveness:

$$
f_1 = \frac{1}{\pi\mu_0\sigma d} < f < \frac{1}{\pi\mu_0\mu_r\sigma d^2} = f_2 \Rightarrow SE_H
$$

$$
= 20\log\left(1 + \frac{2\pi f \mu_0\sigma r d}{3}\right)
$$

$$(7.6)$$

Figure 7.11 shows two end plates of an enclosure which are orthogonal to an incident magnetic field. This field will induce *eddy currents* in the plates. These eddy currents in turn generate a second magnetic field such that the net field through the plates is reduced leading to a reduction of the net magnetic field between both plates. The net field outside the plates is increased.

In the previous case the field was orthogonal to the plates. Figure 7.12 shows the situation where the field is parallel to the plates. Now the eddy currents flow through the loop formed by the four enclosure walls, which are mounted in a tubelike structure. Again the net magnetic field inside the enclosure is reduced and on the outside it is increased.

An alternative formulation of this shielding mechanism is called the *circuit approach to shielding* (Miller and Bridges 1968) and may be more intuitive to understand. For a circular

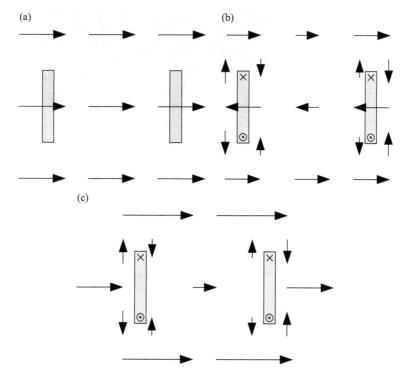

FIGURE 7.11   Eddy currents induced in enclosure plates reduce the field between the plates: (a) orthogonal incident field inducing eddy currents; (b) secondary field caused by eddy currents; (c) net field reduction between two plates.

tube, with radius r, resistance R and self-inductance L, encountered by an eddy current flowing over the circumference of a tube with length l, respectively, are as follows:

$$R = \frac{2\pi r}{\sigma l d} \tag{7.7}$$

$$L = \mu_0 \frac{\pi r^2}{l} \tag{7.8}$$

The lower frequency f_1 at which the eddy current can start to flow is given by

$$R = 2\pi f_1 L \Leftrightarrow f_1 = \frac{R}{2\pi L} = \frac{1}{\pi \mu_0 \sigma r d} \text{ (Hz)} \tag{7.9}$$

When πr is substituted by w, the same crossover frequency as for the flat plate in Figure 4.15a/b is found (with w being the half width of a plate).

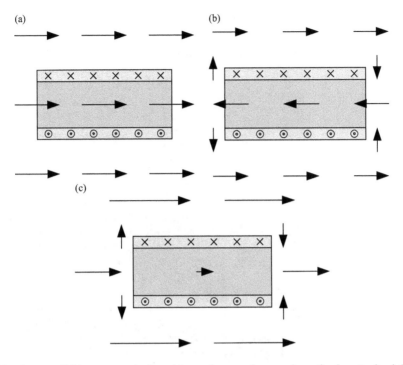

FIGURE 7.12 Eddy currents induced in enclosure plates reduce the longitudinal field between the plates: (a) incident field inducing eddy currents; (b) secondary field caused by eddy currents; (c) net field reduction between two plates.

In the low-frequency domain, the shielding effectiveness increases linearly with frequency as is shown by Eq. 7.6. Furthermore, the shielding effectiveness increases with the dimensions of the cabinet, which is opposite to the findings in the magnetostatic domain.

Also here not always completely closed shielding is required. Figure 7.13 shows a tram passing beneath a passage way between two parts of a hospital. Even when DC traction is applied the current will vary over time due to rectifier ripple and changes in the load condition when trams pass (van Helvoort 2017). The time-varying field related to this current that it could distort equipment on the floor above. As mitigation aluminum plating was placed between the tram traction lines and the passage way (Wouters et al. 2000).

7.2.1.3 Skin effect When the skin effect becomes smaller than the wall thickness of the enclosure (i.e., $\delta < d$), the term kd in Eq. 7.2 becomes larger than 1 and $\cosh(kd) \approx \sinh(kd) \approx \frac{1}{2}e^{kd}$. The magnetic shielding effectiveness then is given by

FIGURE 7.13 Aluminum plates prevent interference between tram traction currents and equipment in the passage way above via eddy current shielding.

$$f_2 = \frac{1}{\pi\mu_0\mu_r\sigma d^2} < f < f_3 \Rightarrow SE_H = 20\log\left(\frac{r}{3\sqrt{2}\mu_r\delta}\,e^{\frac{d}{\delta}}\right)$$

$$= 20\log\left(\frac{r}{3\sqrt{2}\mu_r\delta}\right) + 20\log\left(e^{\frac{d}{\delta}}\right) = 20\log\left(\frac{r}{3\sqrt{2}\mu_r\delta}\right)$$

$$+ \frac{d}{\delta}20\log(e) = 20\log\left(\frac{r}{3\sqrt{2}\mu_r\delta}\right) + 8.686\frac{d}{\delta} \qquad (7.10)$$

In this frequency domain, the current distribution over the wall thickness is no longer homogeneous as shown in Figure 4.2b and the shielding effectiveness increases exponentially with the frequency squared. A thin foil, as shown in Figure 7.14, is sufficient to achieve good shielding. Theoretically, the shielding effectiveness becomes infinite, but in reality will be limited due to technical imperfections of the enclosure. Figure 7.15

FIGURE 7.14 Application of thin copper foil providing skin effect-based shielding for the RF fields employed in MRI.

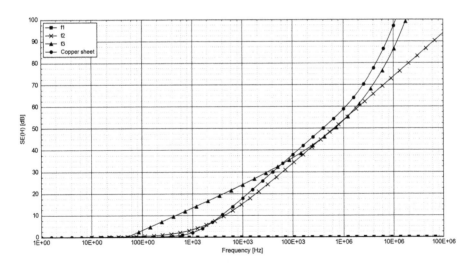

FIGURE 7.15 Magnetic shielding effectiveness for copper foil sphere with a radius of 5 cm and a wall thickness of 0.1 mm calculated with the different approximations: Eqs. 7.6 (square), 7.9 (cross), and 7.10 (triangle) with the full expression given in Eq. 7.2 (bullet).

compares the approximations from Eqs. 7.6, 7.9, and 7.10 with the full expression given in Eq. 7.2.

7.2.2 Impedance concept and transmission line theory

King (1933) and Kaden (1950) calculated the shielding effectiveness by calculating the field at the center of a sphere, caused by a field external to the shield. Schelkunoff (1938, 1943) studied the reciprocal situation where a small current loop is placed at the center of a thin-walled sphere ($r \gg d$), see Figure 7.16. His approach uses a transmission line (see Section 3.4.3), analogy to electromagnetic waves. A shield then can be seen as a change in impedance causing reflections and the shielding effectiveness $SE_{H,S}$ becomes the ratio of the incident wave to the transmitted wave. This can be expressed with three coefficients:

$$SE_{H,S} = R + A + B \text{ (dB)} \tag{7.11}$$

This equation is also known as the *RAB equation*. Figure 7.17 gives a graphical representation of *R, A,* and *B*:

- *Reflection (R)*: the amount of power reflected back to the source. Rather confusingly this term includes the reflection of the incident wave at surface 1 and the reflection at surface 2 transmitted through surface 1.

- *Absorption (A)*: the amount of power absorbed by the dissipative shield during the first pass.

- *Re-reflection (B)*: part of the power which is reflected at the surface 2 is not transmitted back to the source but re-reflected to the shield. Mostly, this term can be ignored due to the large attenuation at higher frequencies.

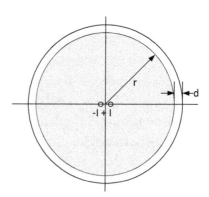

FIGURE 7.16   Current loop shielded by a sphere.

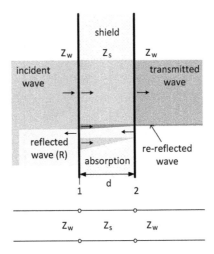

FIGURE 7.17 Transmission line analogy for shielding.

For high frequencies, the reflection coefficient (R) and absorption coefficient (A) of the current source in a sphere can be written as (Schelkunoff 1943):

$$R = 20\log\left(\frac{1}{3|k|}\right) = 20\log\left(\frac{r\sqrt{2}}{3\delta\mu_r}\right) \tag{7.12}$$

$$A = 20\log\left(e^{\frac{d}{\delta}}\right) = \frac{d}{\delta}20\log(e) = 8.686\frac{d}{\delta} \tag{7.13}$$

$$B \simeq 0 \tag{7.14}$$

The total shielding effectiveness then is given by

$$SE_{H,S} = 20\log\left(\frac{r\sqrt{2}}{3\delta\mu_r}\right) + 8.686\frac{d}{\delta} \tag{7.15}$$

The shielding effectiveness $SH_{E,S}$ found in Eq. 7.15 is 6 dB larger than SH_H given in Eq. 7.10. In practice, both are only approximations for the real enclosure, so sufficient margin should be taken into account.

Schelkunoff's approach to shielding effectiveness is widespread in the EMC community. It is easily understood at first glance, in particular in the representation of Figure 7.18 which shows a wave oblique to an infinitely long shield. This picture is easily compared to a light beam bouncing off a mirror.

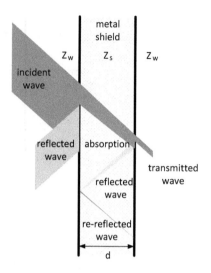

FIGURE 7.18 RAB-model illustrating absorption and re-reflection more clearly, but also leading to confusing with respect to actual physics at lower frequencies.

However, it represents reality only at very high frequency, when the wavelength is only a fraction of the enclosure dimensions. In his approach, Schelkunoff postulated that the reflected wave travels along the same path as the incident wave (as indeed is the case in a transmission line). In most practical situations, *diffraction* of the wave around the enclosure will take place, which can be understood as a secondary field generated by the currents induced in the three-dimensional shield by the incident field (see Figures 7.11 and 7.12).

Most often, the RAB equation for an infinite plane is encountered:

$$SE_S = R + A + B \text{ (dB)} \tag{7.16}$$

$$R = 20\log\left(\frac{|k+1|^2}{4|k|}\right) \tag{7.17}$$

$$A = 20\log\left(e^{\frac{d}{\delta}}\right) = \frac{d}{\delta}20\log(e) = 8.686\frac{d}{\delta} \tag{7.18}$$

$$B = 20\log\left|1 - \left(\frac{k-1}{k+1}\right)^2 e^{-2\frac{d}{\delta}}\right| \tag{7.19}$$

$$k = \frac{Z_w}{Z_s}, \text{ with } Z = \sqrt{\frac{j\omega\mu_0\mu_r}{\sigma + j\omega\varepsilon_0\varepsilon_r}} \Rightarrow Z_w = \sqrt{\frac{\mu_0}{\varepsilon_0}} = 377\Omega,$$

$$Z_s = \sqrt{\frac{j\omega\mu_0\mu_r}{\sigma}}$$

(7.20)

For high frequencies, the RAB calculation for an infinite plane is only 2.5 dB lower than for the sphere, because the term B can be ignored, the expressions for R are nearly identical, and A is identical in both cases.

7.2.3 **Multilayer shields**

For low frequencies, e.g., AC power, a sufficiently large shielding effectiveness may lead to an unfeasible thick shield. In this situation using multiple layers may offer a solution, because they can obtain higher shielding effectiveness as single layer with comparable total thickness.

7.2.3.1 ***Combining ferromagnetic and nonmagnetic shields*** For enclosures both Kaden (1950) and Schelkunoff (1943) advise a combination of a copper (nonmagnetic, nm) shield and a ferromagnetic (fm) shield (Figure 7.19). Maximum attenuation is obtained if the ferromagnetic layer is kept closest to the disturbance source. Assuming that no skin effect occurs in either shield ($d_{nm} < \delta_{nm}$, $d_{fm} < \delta_{fm}$) for a source inside the sphere it is found (Kaden 1950):

$$SE_H = 20\log\left(\frac{2\pi f\mu_0\mu_{r,fm}\sigma_{nm}d_{nm}d_{fm}}{3}\right),$$

(7.21)

ferromagnetic layer on the outside

$$SE_H = 20\log\left(\frac{4\pi f\mu_0\mu_{r,fm}\sigma_{nm}d_{nm}d_{fm}}{3}\right),$$

(7.22)

ferromagnetic layer on the inside

These equations resemble Eq. 7.6, but they combine the highest conductivity with the highest permeability. Furthermore, the term rd has been replaced by the term $d_{nm}d_{fm}$, making the shielding independent of the enclosure dimensions. For maximum SE_H, the thickness of both shields has to be equal:

$$d_{nm} = d_{fm} = \frac{1}{2}d.$$

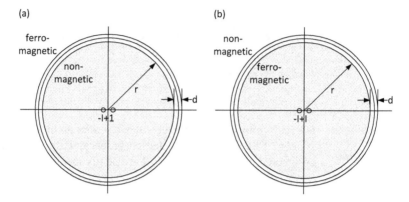

FIGURE 7.19   Current loop shielded by a double-layer sphere: (a) ferromagnetic on the outside; (b) ferromagnetic on the inside.

7.2.3.2 Multilayer shields in open structures   As seen in Chapter 6, open magnetic structures often show field enhancement at their edges, which may limit their application; furthermore, outdoors (as shown in the case study of Figure 7.13) aluminum is more easy to install (weight) and maintain (corrosion protection). Figure 7.20 shows another case where a power transformer is placed in the basement of a residential building (Wouters et al. 2000). The current paths were well defined and no large common mode currents were present, still the magnetic field in the apartment above the basement reached values as high as $100\,\mu T$. Simulations (Figure 7.20b) showed a reduction of only a factor of 6.4 for a 10-mm-thick aluminum plate with a width of $2\,m$. As alternative a combination of two shields was applied: a plate together with a tray at $20\,cm$ above the conductors. Both were made from 3-mm-thick aluminum. With this multilayer approach a factor of 15 in reduction was realized with less material. The resulting field distribution is shown in Figure 7.20c.

7.2.4   Slits, slots, and holes

Enclosures normally have a lid or a door. Figure 7.21 shows a setup to demonstrate the effect of the slits between a box and its lid. It consists of two large coils, which make a magnetic disturbance field in the center between them (similar to *Helmholtz coils*, which will be discussed in Section 7.3). These coils create a 50-kHz magnetic field, which induces $30\,mV$ in an unshielded pickup coil. In addition, it employs a 2-mm-thick aluminum box with identical dimensions as in Figure 7.8.

Theoretically, the attenuation of the box at 50 kHz is 77 dB. The voltage measured with the sensor inside the box, therefore,

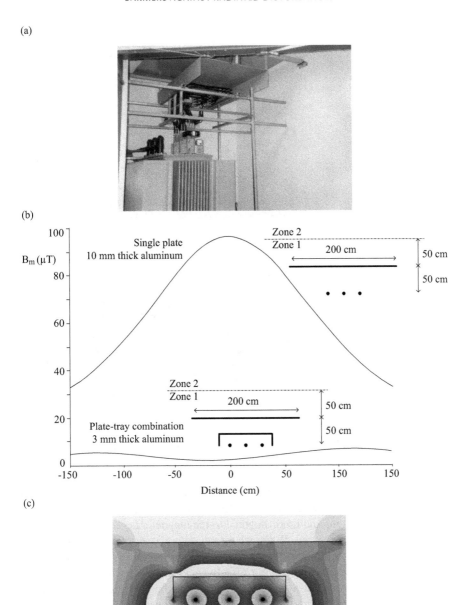

FIGURE 7.20   Two aluminum layers shield top floor from fields caused by power cables: (a) picture; (b) simulation results for single and dual layer; (c) simulated magnetic field. Dark colors identify a large field.

(a)

(b)

FIGURE 7.21 Use of (a) Setup and (b) test box to demonstrate the influence of slits on the shielding effectiveness.

should drop with a factor of 7000 to 4 μV, which is too low for the oscilloscope to detect. Indeed, if we place the box with bottom and lid orthogonal to the magnetic field, no voltage can be measured. When we rotate the box such that the lid is parallel to the field the measured voltage becomes 7 mV. This means that the shielding effectiveness has dropped to 12.6 dB. The root cause for this degradation is the nonconductive aluminum oxide preventing the current flow over the box to close via the lid (see Figure 7.22). Steel enclosure may show the same issue, because often they are painted before being assembled preventing proper electrical contact over the full circumference.

If only limited shielding is required the box may remain open. If the lid is removed in the situation from Figure 7.22a, the measured voltage becomes only 1 mV. In the situation of Figure 7.22b, it becomes 15 V and hardly any attenuation remains. If the magnetic field is perpendicular, a single

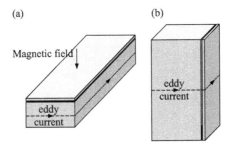

FIGURE 7.22 Current flow over a box with a lid: (a) current has not to flow over the slit between the box and the lid; (b) current flow is hampered by slit.

aluminum or steel plate may be sufficient as illustrated by the applications in Figures 7.13 and 7.20.

7.2.4.1 Seams In case more shielding is needed the box has to be closed and the maximum shielding efficiency in Figure 7.7 between f_3 and f_4 is determined by the amount of leakage through slits, slots, and holes. A reasonable obtainable value is 60 dB, with some care 100 dB or more can be achieved. In the latter case, metal parts have to be welded together or overlap must be ensured. If sufficient overlap is present no *gaskets* are required. Some examples are shown in Figure 7.23. For a high-quality connection between aluminum parts, the construction on the left-hand side of Figure 7.23c is advised. The connection should be made very tight such that the gasket frets through the oxide layer and simultaneously creates an airtight seal so that no new oxide layer can grow.

7.2.4.2 Holes at low frequency For a single *hole* with radius r_h in a sphere with radius r_s holes, an exact derivation has been obtained for the shielding effectiveness in the center of the sphere by Kaden (1950):

$$SE_H \approx 20\log_{10}\left(\frac{\pi r_s^3}{r_h^3}\right) \text{ (dB)} \tag{7.23}$$

This approximation is valid as long as the diameter of the shield is smaller than the wavelength and the shield is infinitely thin. For realistic shields with a thickness d, additional attenuation is obtained (Kaden 1950):

$$SE_H \approx 16\frac{d}{r_h} \text{ (dB)} \tag{7.24}$$

(a)   (b)

(c)   (d)

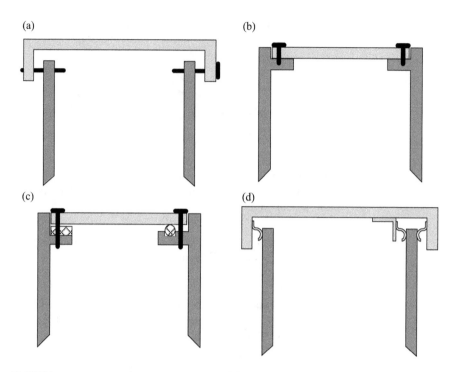

FIGURE 7.23 Proper designs minimize field penetration through a *seam*: (a) overlap; (b) improved overlap; (c) spring feathers; (d) gaskets.

The attenuation can be further increased by deliberately mounting a tube over the hole (see Figure 7.24). The shielding effectiveness can be calculated by replacing the thickness d in Eq. 7.24 by the length of the *waveguide l*. Such a construction is often referred to as a *waveguide below cut-off* (WBCO), where the cutoff frequency is given by Harvey (1963):

$$f_C \approx 1.8412 \frac{c}{2\pi r_h} = \frac{8.8 \cdot 10^7}{r_h} \text{ (Hz)} \qquad (7.25)$$

where c is the speed of light in air. A practical implementation of multiple holes and tubes is a *honeycomb* as shown in Figure 7.25. During installation special attention has to be given to the mounting in order to prevent long slits around the circumference of the honeycomb. Figure 7.26 shows that it is better to have multiple small holes instead of one large hole, because the field at a given position does not increase linearly with the number of holes. In addition, small holes allow shielding up to higher frequencies.

FIGURE 7.24 Conductive tubes strongly enhance the attenuation of holes in a shielding enclosure.

FIGURE 7.25 Honeycomb typically function as air vents.

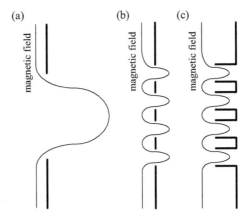

FIGURE 7.26 Holes reduce the shielding effectiveness of an enclosure: (a) large hole; (b) small holes; (c) holes with tubes. The shieling effectiveness increases from left to right.

7.2.4.3 *Holes at high frequency*   When the wavelength has the same order of magnitude as the aperture a common rule of thumb is given by Ott (2009):

$$SE_{Aperture} \approx k \log_{10} \frac{\lambda}{2L} \qquad (7.26)$$

where $k = 20$ for a rectangular slot and 40 for a round hole. L is the largest dimension of the slot, or the diameter of the round hole. In case multiple apertures are used within a distance of $\lambda/2$ of each other, Eq. 10.23 is corrected with the number (N) of holes:

$$SE_{Aperture} \approx k \log_{10} \frac{\lambda}{2L} - 10 \log_{10}(N) \qquad (7.27)$$

When shielding is applied to attenuate too large emissions from a disturbance source, the assesment if compliance is reached is typically specified as a limit for the electric field E_{far} at 10 m distance. For a rectangular slot with the length L and the width W and a magnetic field H_{wall} at its inside wall, this far field is given by Li et al. (2001) and Harberts et al. (2013):

$$|E_{far}| = 8.8 \times 10^{-17} \frac{4\pi^2 f^2 L^3 |H_{wall}|}{|R \ln(1 + 0.66 W/L)|} \qquad (7.28)$$

where R represents the measurement distance in meters

7.2.5 Resonances

Figure 10.7 shows a reduction in shielding effectiveness for frequencies above f_4. At these frequencies the wavelength becomes comparable to the enclosure size, which will act as resonant cavity. For a rectangular enclosure with dimensions $h \times b \times w$ (m) resonances will occur at

$$f_{i,j,k} = \frac{3 \times 10^8}{2} \sqrt{\left(\frac{i}{h}\right)^2 + \left(\frac{j}{b}\right)^2 + \left(\frac{k}{w}\right)^2} \qquad (7.29)$$

The indices i, j, k are integers ranging from zero to infinity, but at least one is nonzero. The lowest resonance frequency f_4 is given by

$$f_4 = \frac{3 \times 10^8}{2} \sqrt{\left(\frac{1}{l}\right)^2} \qquad (7.30)$$

where l is the largest dimension of the enclosure. For the box of Figure 7.21, a first resonance is found at 2 GHz. When the dimensions increase the resonance frequencies are

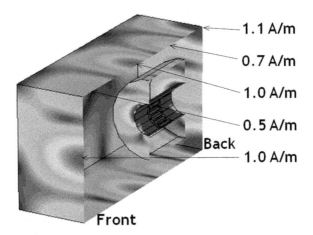

1.1 A/m

0.7 A/m

1.0 A/m

0.5 A/m

Back

1.0 A/m

Front

FIGURE 7.27 Amplitude plot of magnetic field strength showing resonance in an MRI examination room. Dark colors indicate high density.

lowered. Figure 7.27 shows resonances in the examination room (5.3 × 3.4 × 2.9 m) of an MRI scanner at its operating frequency of 128 MHz (Harberts et al. 2013).

In practical situations, however, enclosures are not empty, but contain other equipment or electronics. In effect, this will alter (increase) resonance frequency and provide extra damping of the fields inside the enclosure (Worm and Kanters 2004). Another way to reduce the standing waves that are caused by resonance is to add attenuation by adding self-adhesive sheets of ferrite material.

7.3 Active protection

Magnetic (and electric) fields can be superimposed. The total net field is the sum of the individual fields. For low frequencies (roughly between DC and 5 kHz) this can be used for creating a field-free region, by purposely creating a field which counteracts the incident field as shown in Figure 7.28. Most often *Helmholtz coils* (Figure 7.29) are used, because they create a homogeneous field in a rather large volume in the center between the coils (Ruark and Peters 1926). The field in the center is oriented along the dash-dot axis. Its value at the center point is given by

$$H_{Helmholtz} = \frac{8}{\sqrt{125}} \frac{I \times N}{r}$$

$$(7.31)$$

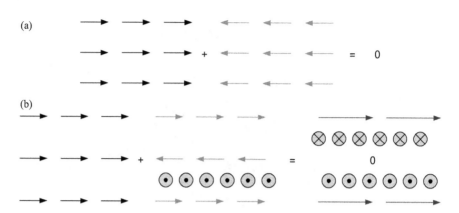

FIGURE 7.28 Active protection can be used to create field-free regions at extremely low frequencies without magnetostatic shielding: (a) fields can cancel each other out; (b) active coils used to create a field-free region.

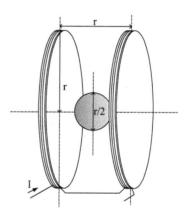

FIGURE 7.29 A Helmholtz coil configuration consists of two coils with equal current flow, thus creating a homogeneous field in the center between the coils along the axis of the coil pair.

where I is the current through the coils, N the number of windings per coil, and r both the radius and the distance between the coils. As a rule of thumb, the field can be considered to be homogeneous in a volume of $r/2$ (ASTM A698 2002). Multiple Helmholtz coils can be combined to control the magnetic field over multiple axes.

Active shielding can also be used to reduce the fringe field of a magnet. An example is given in Figure 7.30. The disadvantages of active shielding are the required field detection sensors and the addition of control electronics. Another disadvantage

(a)

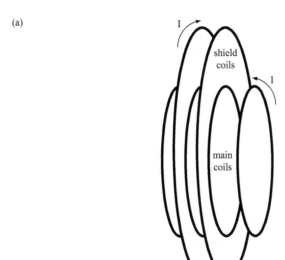

(b)

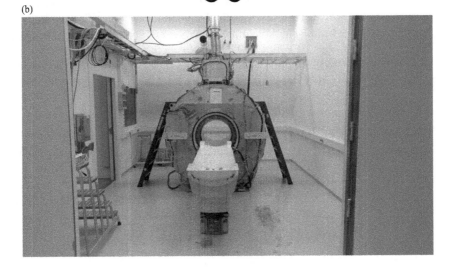

FIGURE 7.30 Active protection is used to reduce the fringe field of superconducting MRI magnets: (a) schematic of windings; (b) picture of fully assembled magnet.

may be the increase of field strength near the active coils on the outside of the protected area.

7.4 Test methods

There are a variety of standardized test methods for performing pre-compliance shielding tests of enclosures and large shielded rooms and components used in such enclosures.

7.4.1 Large enclosures or shielded rooms

The most common method for testing the shielding effectiveness of large enclosures was MIL-STD-285 (1965) and is still often referred to. This standard, however, was extended and superseded by International Electrical and Electronics Engineers Standard 299 (IEEE Std 299 2006) in 1997. It can be applied to enclosures which are larger than 2 m. For low frequencies, between 9 kHz and 20 MHz (although downscaling to 50 or 60 Hz is possible), the magnetic shielding is tested by placing current loops on opposite site of the shield as shown in Figure 7.31. The loops have to be repositioned around the enclosure to ensure that weak spots in the shielding are identified.

For frequencies between 20 and 100 MHz, biconical antennas are used and again these have to be repositioned in the enclosure and both horizontal and vertical orientation of the transmitting antenna (on the outside) has to be used (see Figure 7.32). The IEEE tests provide near-field measurement results while final compliance is specified as far-field limit. The shielding effectiveness determined according to the IEEE Std 299 near-field method agrees accurately with the reduction of the far-field emission determined according to IEC 60601-1-2 (IEC 60601-1-2 2014; IEEE Std. 299 2006; Harberts et al. 2013).

7.4.2 Cabinets and sub-racks

The International Electrotechnical Commission (IEC 61587-3 2013) describes an electromagnetic shielding performance test for empty *cabinet*s and sub-racks. It uses a small battery powered disturbance source, called a *spherical dipole antenna* (SDA), which is controlled via a fiber optic link. The SDA is placed inside the rack and the disturbance is measured at 3 meter distance as shown in Figure 7.33.

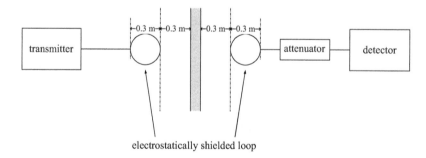

FIGURE 7.31   Two loops are used to test shielding effectiveness.

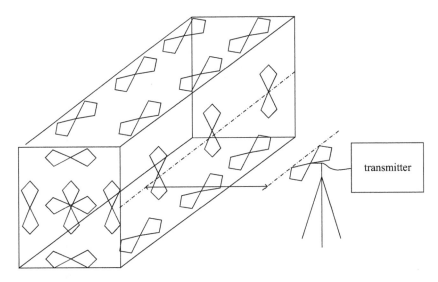

FIGURE 7.32   Two biconical antennas are used to test shielding effectiveness. The receiving antenna is placed inside the shielding enclosure at various positions and orientations (horizontal and vertical). The distance between wall and antenna has to be at least 0.3 m; antenna inside the shielding enclosure is moved around. The transmitter is placed on a wooden table (insulated support).

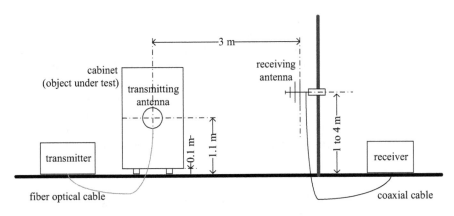

FIGURE 7.33   Spherical dipole antenna setup for testing empty cabinets and sub-racks.

7.4.3 Small enclosures

IEC 61000-5-7 (2001) parts 5–7 provide performance requirements, test methods, and classification procedures for the shielding performance of empty enclosures in the frequency range between 10 kHz and 40 GHz with the setup shown in Figure 7.34. The *shielding performance (SE)* is determined

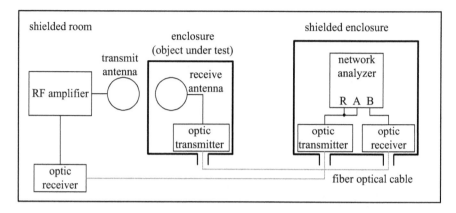

FIGURE 7.34   Enclosure testing according to IEC61000-5-7.   The network analyzer has an RF output (*R*) which is connected via fiber optics to the transmit antenna and measured on input port (*A*). The signal detected by the receive antenna is measured on input port (*B*). The fiber optic cables leave the shielded enclosures via waveguides as has been discussed in Section 7.2.4.

by measuring the power transfer between both antennas in two sweeps, one with the enclosure under test present and one without. For low frequencies, loop antennas are used; for mid frequencies, monopoles or dipoles are used; and for high frequencies, horn antennas are required.

The results are classified according to Tables 7.2 and 7.3, and the enclosure can be labeled as EM*ABCDEF*. In case a narrow band source radiating at 100 MHz needs 60 dB shielding, the specification of the enclosure will become EMxxx6xxx.

IEEE Std 299 (2006) has been extended with IEEE 299.1 (2013) to include higher bandwidths and enclosures smaller than 2 meters (down to 10 cm). It includes the use of reverberation rooms and is less applicable to large installations and

Table 7.2   EM shielding code—shielding designators

Frequency band	Shielding designator
10–100 kHz	A
100 kHz–1 MHz	B
1–30 MHz	C
30 MHz–1 GHz	D
1 GHz–10 GHz	E
10–40 GHz	F

Table 7.3 EM shielding code—shielding designator value

Shielding performance (dB)	Shielding designator value
Untested	X
<10	0
≥10	1
≥20	2
≥30	3
≥40	4
≥50	5
≥60	6
≥70	7
≥80	8
≥90	9
≥100	10

complex systems. Similarly, dedicated setups to test the shielding performance of materials and gaskets on component level, such as IEEE 1302 (2008) and ASTM D4935-10, are beyond the scope of this book.

CHAPTER EIGHT

Documenting electromagnetic compliance

The previous chapters, except Chapter 1, had a strong focus on the technical measures which helped to achieve electromagnetic compliance. As systems and installations are becoming increasingly complex, it is of increasing importance to prove that the design is correct in a traceable and verifiable way. Making the right product that fulfills standards and regulations is critical and customers have become more and more aware of the financial effects for not meeting electromagnetic compatibility (EMC) requirements and request for documentation.

In consequence, EMC has to be embedded in the organizational process, which is typically a project organization. Systems engineering will be used as a tool to map EMC activities on the various project phases. Next, the role of standardization and regulation is discussed followed by a framework for documenting conformance to regulation.

8.1 Project management plan

When designing and building a complex installation or large system, the *project management plan* contains multiple aspects, such as scope, requirements, risk, procurement, and quality (PMI 2013). Ideally, a dedicated EMC management plan is added to the overall project management plan that covers the following:

- The general EMC requirements of the installation or system

- The individual EMC requirements of modules or subsystems
- The electromagnetic environment in which the system or installation needs to function
- Other electromagnetic requirements, such as safety and exposure of persons to electromagnetic fields
- EMC requirements during purchasing
- Assuring EM compliance after retrofits, updates, or upgrades
- EMC maintenance during project life cycle
- Documenting requirements, analysis, tests, and the responsible persons

The above list shows no particular order related to importance or chronology; however, documentation is extremely important to prove compliance to regulatory bodies or as evidence in case of dispute.

8.1.1 V-model

Properly managed projects for engineering large installations and complex systems employ the *V-model* (Forsberg and Mooz 1991) (see Figure 8.1). The downward slope shows the engineering process: stakeholder requirements are translated to technical requirements on system (installation) level and then decomposed to requirements on the subsystem and unit level.

At the bottom part, the requirements are decomposed to the level that units actually can be constructed and physically being built (or implemented).

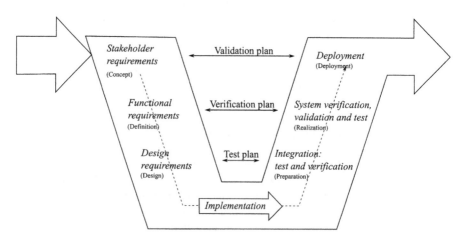

FIGURE 8.1   The V-model describes the various project phases in systems engineering.

The rising slope shows the creation process: units are integrated, tested, and verified. Testing, verification, and validation are performed against the requirements generated in the downward slope. This is indicated by the horizontal arrows between the slopes.

The V-model emphasized the need for testability of the design, meaning that a requirement is only valid if it is accompanied by a test methodology. Therefore, it is a very useful model to design-in electromagnetic compliance. High-level requirements (typically demanded by regulations) are decomposed to design or procurement criteria for units, and measurements are made on all levels such that nonconformance can be identified early via the *IVVQ* process:

- *Integration*: The process of assembling all parts and components to construct the system or installation according to guidelines.

- *Verification*: The process of determining if functions and requirements are correctly implemented. Alternatively phrased, verification is the answer to the question if the system or installation is implemented the right way.

- *Validation*: the process of determining if the customer's needs are satisfied. Validation is then the answer to the question if the right system or installation is implemented.

- *Qualification*: the process of demonstrating that the specified requirements are met. Qualification, therefore, answers the questions how tests prove that the system or installation meets the requirements.

8.1.2 Life cycle management

Complex systems and installations have a long life cycle and evolve over time. New features are added, obsolete components are replaced and the product's abilities are pushed toward its limits. These changes affect the EMC behavior of the system or installation both in its functional aspects as well as its legal aspects. Keeping track of changes and their effect is mandatory in the process of maintaining EMC, which extends way beyond the initial project and therefore the proper process has to be defined by the project as deliverable.

Many systems and installations undergo mid-life upgrades as their economic life span exceeds their technical life span in many occasions. New functionalities might be required by the customer, hence new requirements surface. The result is a new project that involves systems engineering. This puts additional importance on thorough documentation of all processes. For EMC activities it is important to know that these activities are also included in the deployment and maintenance phase. If equipment has to be replaced, it might be necessary to reinvestigate the EMC behavior of the entire system or installation.

8.2 Standardization and regulation

In a world with ever-increasing complexity it is needed to create some way of formalizing common practices or agreed ways of working, which are acceptable by manufacturers and installers, customers, and *regulators*. Standardization organizations, like IEC and ISO, facilitate this process by initiating and maintaining standards (ISO/IEC 1996):

STANDARD

a document, established by consensus and approved by a recognized body, that provides for common and repeated use, rules, guidelines or characteristics for activities or their results, aimed at the achievement of the optimum degree of order in a given context.

Often standards are adopted by regulatory organizations and thereby become enforced by law.

8.2.1 International standardization bodies

International standards are developed either by the *International Standardization Organization* (ISO) and the *International Electrotechnical Commission* (IEC). The international EMC standards are developed by the IEC where national standardization bodies participate in the development of these standards.

8.2.2 National standardization bodies

National standardization bodies adapt international to *national standards* and/or develop specific national standards. When adopted these standards may be used for governmental regulation, making them mandatory. Often different bodies are responsible for civil and military standards or have a specific focus area: Germany has the VDE for civil systems and VG for military systems as the French use their GAM for military standards. The British civil standard is BS and their military standard is DEF-STAN. The Society of Automotive Engineers (SAE) provides rules and regulations for automotive EMC. Fortunately, many of these standards share the same methodologies and are more and more based on the international standards.

In EMC the U.S. institutes such as American National Standards Institute (ANSI) and Federal Communications Council (FCC) play an above average role.

8.2.2.1 American National Standards Institute The *ANSI* is the main standards body in the United States. Most of the processes are incorporated in other standards that might exist. Within ANSI, the Accredited Standards Committee (ASC) C63 coordinates the EMC activities. It consists of subcommittees (listed in Table 8.1) to which the European EMC standards show resemblance.

8.2.2.2 Federal Communications Council The *FCC* (*Federal Communications Council*) is charged with the regulation of radio and wire communication. The Code of Federal Regulations contains the following:

- Title 47, part 15: Radio Frequency devices
- Title 47, part 18: Industrial, Scientific and Medical Equipment

Both parts contain limits on emission, where FCC title 47 applies to telecommunication equipment in general and part 15 applies to radio frequency devices in particular. FCC distinguishes intentional radiators from unintentional radiators.

- *Intentional radiators* are designed with the intention to emit RF energy, i.e., emission of RF energy is part of their function. Part 15 refers to low power transmitters, such as door openers, baby monitors, wireless home security systems. They have in common that the antenna is either integrated as a part of the product or the connection between the equipment and the antenna is done using a dedicated connector.
- *Unintentional radiators* refer to equipment that generates RF energy to be able to operate but not intended to emit

Table 8.1 . ANSI C63 (EMC) subcommittees

Subcommittee	Title
1	Measurement techniques and developments
2	EMC definitions
3	International standardization
5	Immunity
6	Test laboratory and accreditation

RF energy like transmitters. Examples of this type of equipment are radio receivers and detectors (with built-in oscillators), household equipment, (switch mode) power supplies, and so on.

Unintentional radiators are grouped in two categories:

- *Class A*: for use in industrial, commercial, or business environment.

    This kind of equipment is not intended to be used at home. Equipment that is categorized as Class A equipment shall carry a label that states that it concerns equipment that is not intended for domestic use.

- *Class B*: for use in a residential environment.

    This kind of equipment is marketed for use by the general public. Examples are television sets, pocket calculators, and personal computers.

This grouping differs from the IEC 61000-6-4 (2018) zone classification, which was discussed in Section 1.3.2.

The following approval methods can be applied, depending on the product:

- *Certification*: Products must be tested and a technical file must be submitted to and approved by the FCC.

- *Verification*: Products must be tested and the records are kept by the manufacturer, however, no file is submitted to nor approved by FFC.

- *Declaration of conformity (DoC)*: Testing is performed by laboratories which have been certified by FCC-recognized accreditation bodies. It is in common use for class B computers and peripherals.

8.2.3 **Regional standardization bodies**   Within the European Union, the *European Committee for Electrotechnical Standardization (CENELEC)* is responsible for developing harmonized standards which refer to the IEC standards, like the IEC-EN 61000 series. These are useful to show compliance to the European EMC directive. The other two bodies in Europe are the *European Committee for Standardization (CEN)* and the *European Telecommunications Standards Institute (ETSI)*. Often they work together in joint groups.

8.2.3.1 European EMC directive   The European EMC directive harmonizes the laws on electromagnetic compatibility in the EU Member States and was first introduced in 1992. It received updates in 2004 and 2014. It defines the basic legal requirements, called essential requirements, and leaves technical details to the European standardization bodies listed above. The *essential requirements* are as follows (Directive 2014/30/EU 2014):

1. *General requirements*

   Equipment shall be designed and manufactured to ensure that

   a. The electromagnetic disturbance generated does not exceed the level above which radio and telecommunications equipment or other equipment cannot operate as intended.

   b. It has a level of immunity to the electromagnetic disturbance to be expected in its *intended use* which allows it to operate without unacceptable degradation of its intended use.

      (Implicitly this also means that equipment shall not interfere with itself. In particular, for complex installations and large systems this is not trivial due to the number of electronic subsystems and modules involved.)

2. Specific requirements for fixed installations

   A fixed installation shall be installed applying good engineering practices and respecting the information on the intended use of its components, with a view to meet the general requirements (item 1) as described before.

The EMC directive only applies to equipment which is available on the market for the end-user. It does not include subsystems or units which are intended to be incorporated in a larger system or fixed installation.

In the first case, the manufacturer of the large system will have to prove compliance of the integrated system with the EMC directive. This proof has to include an electromagnetic compatibility assessment and *technical documentation* containing the following:

1. General description of the system.

2. Conceptual design and manufacturing drawings, schemes of components, etc.

3. Descriptions and explanations for the above mentioned drawings.

4. List of harmonized standards used in the EMC assessment. In case these standards not have been applied an

additional description is needed to show how the essential requirements are met.

5. Results of calculations, examinations, etc.

6. Test reports.

This technical document is part of the *Technical Construction File (TCF)*. The TCF may include additional design and test information required by other regulations such as the Machine Directive (Directive 2006/42/EC 2016).

In the second case, the EMC directive mandates that the good engineering practices referred to by the essential requirements under point 2 have to be documented. Again, this document is included in the *TCF* and may include additional information mandated by other legislation. The TCF has to be held by the responsible person(s) for as long as the fixed installation is in operation. Upon request it has to be presented to the national authorities for inspection.

8.2.3.2 CE mark   The *CE Mark* enables free trade among all European countries without restrictions, because it implies conformance to European regulation. CE stands for "Conformiteé Europeenne", which is French for European conformance. Nevertheless, when using CE Marked equipment care must be taken that the environment intended by the manufacturer matches the environment of the user as has been discussed in Section 1.2.2.

8.2.4 **Professional societies**   The most well-known professional societies involved in EMC are the International Electrical and Electronics Engineers (IEEE) in the United States and the Institution of Engineering and Technology IET in the United Kingdom. In particular the IEEE, often in cooperation with ANSI, provides excellent background documents to design EMC for installations, e.g., ANSI/IEEE STD 518 (1996).

8.2.5 **Classification**   An overview of organizations providing standards related to EMC is given in Table 8.2. The standards are classified into eight types (Ogunsola and Mariscotti 2013):

1. *Basic standards*: describe the fundamentals (see Table 8.3).

2. *Generic standards*: They deal with general electromagnetic environments in which equipment has to operate for which no specific standard exist or do not yet exist. If no specific (product) standard exists, one can use the generic standards.

Table 8.2 Classification of standards mapped on standardization organizations

Function origin	Type of standard							
	Basic standards	Generic standards	Product standards	Design standards	Process standards	Specifications and codes	Management system standards	Personnel certification standards
FCC		✓	✓		EMC process standards do not exist at the time of writing			
EEC	✓	✓	✓	✓				
ITU		✓	✓					
IEEE/IEE (IET)	✓	✓	✓	✓		✓		
ANSI	✓	✓	✓				✓	
ETSI	✓	✓	✓			✓		
CENELEC	✓	✓	✓					
ISO	✓	✓					✓	
NARTE								✓
CISPR	✓							

3. *Product standards*: They apply to a specific range of products like Information Technology Equipment (ITE) or the standards for adjustable speed drives to name a few.

4. *Design standards*: They specify how a product is to be constructed. In terms of EMC, a design standard could define how a product is to be constructed in order to comply. The IEC 61000-5-2 could be regarded as a design standard as it provides installation and mitigation guidelines.

5. *Process standards*: As the name suggests, they describe the way of working. What has to be done and in which subsequent order to achieve reproducible results.

6. *Specifications and codes*: Specifications are a set of conditions and requirements that are very specific about, in this case, EMC specifications. These specifications are, e.g., to be used by purchasing departments, enabling them to buy the equipment with the desired or engineered EMC behavior. Codes refer to standards that are declared by law.

7. *Management system standards*: An EMC laboratory is organized in a way that reproducible results are achieved. Examples are the calibration procedure, the test guidelines for the EMC engineers, the way the equipment under test is handled, and so on. The ISO 17025 is the standard that applies to calibration and test laboratories.

8. *Personnel certification standards*: Can be used as proof of available competence in a company. NARTE is an example to certify EMC personnel. Companies might require for example electrostatic discharge (ESD)-certified personnel to operate within their premises.

As their name implies, the IEC's basic EMC publications specify the general conditions or rules necessary for achieving electromagnetic compatibility. They serve also as building blocks for the IEC technical committees that develop EMC product Standards.

The structure of the IEC 61000 as presented in Table 8.3 might appear to be a bit confusing as it not only contains normative standards but also technical reports that are nonnormative such as the guidelines on installation and mitigation and the description of the electromagnetic environment. The parts 7 and 8 are still open. Therefore, most standards fall in multiple categories as shown in Table 8.2, e.g., IEC does not only

Table 8.3 Structure of the IEC 61000

Part	Scope	Description
1	General considerations	Provides general information like basic principles, definitions, and terminology
2	Environment	Describes the electromagnetic environment and classified these electromagnetic environments and their compatibility levels
3	Limits	Provides limits on emission and immunity
4	Testing and measurement techniques	Provides information on test and measurement techniques
5	Installation and mitigation guidelines	Provides installation and mitigation measures
6	Generic standards	These are the generic emission and immunity standards
7	Open	
8	Open	
9	Miscellaneous	

Table 8.4 Structure of CISPR 16 structure (IEC-CISPR 16 2018)

Part	Description
1	Specifies voltage-, current-, and field-measuring apparatus, and test sites. These include calibration and verification aspects of measuring apparatus
2	Specifies the methods for measuring high-frequency EMC phenomena, dealing both with disturbances and immunity
3	An IEC Technical Report that contains specific technical reports and information on the history of CISPR
4	Contains information related to uncertainties, statistics, and limit modeling

provide the basics (category 1) about electromagnetic environments, but also provides aid for development in its 61000-5-2 Installation and mitigation guidelines (category 4). CISPR 16 specifies equipment and methods for measuring disturbances and testing for immunity as shown in Table 8.4.

8.3 EMC documentation framework

Figure 8.2 provides a framework for documenting electromagnetic requirements. It resembles a house (or temple) and consists of the following:

- *EMC management plan* (introduced in Section 8.1)

    The EMC management plan describes how all relevant stakeholders will have the right level of EMC knowledge at the right time. This includes two aspects:

    - Planning of specific deliverables by specialists at each project milestone.

    - Creating sufficient EMC awareness throughout the (project) organization. Consider a specialist who purposely designs in an EMC gland, but a purchaser procures a different type for cost reasons or the engineer in the field forgets to remove the insulation from the cable before feeding it through.

- EMC requirements overview

    This document contains the requirements that apply to EMC: Which requirements and how are they described? Are there any other requirements that affect EMC requirements that might originate from the equipment's environment? For instance: if the installation is in the neighborhood of flammable substances, it might be the case that it limits the emission level. Safety critical equipment like transmitters and receivers might also affect the emission frequency and level to avoid interference. The same applies to ESD and sensitive equipment and flammable substances.

- EMC control plan

    The EMC control plan can be a spreadsheet that contains specific information on how compatibility is guaranteed during construction phase as well as maintaining compatibility during the life cycle of the system or installation. For example, if a PLC is defective it has to be replaced. It is important to consider the EMC merits of the replacement PLC to assure continuation of compliance.

- EMC analysis documents

    EMC analyses are used to provide the rationale to EM compliance of the entire system or installation. If for some reason it is deemed necessary to analyze parts of the system, it is documented in the EMC analysis document. It can be done using mathematical analysis, engineering tests or compliance tests. It always concerns parts or modules.

- EMC test plan

    When complex systems and installations are being tested and measured, it is necessary that top-level EMC requirements are "translated" into EMC requirements that apply to modules and submodules such that testing can take place early in the V-model (see Figure 8.1). See Section 8.3.1 for more details.

- EMC test report(s)

    Test reports provide the observed results of the tests as prescribed in the EMC test plan.

    See Section 8.3.2 for more details.

- Final EMC analysis

    The final EMC analysis is a brief analysis that draws a conclusion that the system or installation as a whole will meet the EMC requirements, based on evidence that is provided by the underlying documents: This is the assumption of conformity on which the *Declaration of Conformity* will be based.

8.3.1 **EMC test plan**   An EMC test plan provides a detailed plan that describes the following issues:

- The name of the manufacturer, the type number, and the serial number of the equipment that is tested
- The function of the equipment that is tested
- A description of what is actually tested
    - The power supply: single or three phase, DC, etc.
    - Hardware revision number
    - Software revision number
    - Firmware revision number
    - Settings of the equipment that guarantee maximal emission and/or maximum susceptibility
    - A description of the auxiliary equipment that is used to make the system operational
    - If applicable, a description of equipment status logging that is especially useful to be able to verify correctness or operation during immunity tests
- A map that provides information about how the equipment was physically placed during tests

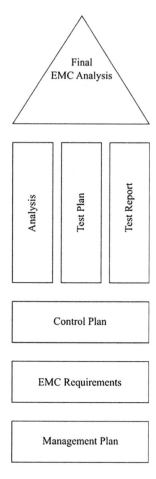

FIGURE 8.2 Framework for documenting electromagnetic compliance.

- An electric block diagram that provides information about the cable types and their lengths as well as how the cables are physically routed
- A description where the measurement results are taken and where the immunity tests are done (especially important for ESD tests)
- Specific information about standards
  - Used limits during emission measurements
  - Used dwell time for each frequency step during immunity tests

- Used immunity test levels
- Used carrier modulation (AM, FM, PM) and modulation frequency during immunity tests
- An explanation and specification how the relation between the overall emission limits are decomposed into module limits
- A description about deviations from the used limits. Reasons can be the following:
  - Another carrier modulation is used because neighboring equipment makes it necessary.
  - Increased immunity test levels because safety equipment is involved or mobile communications equipment can be used in the equipment's vicinity.
  - Extended frequency range (for example) if it is anticipated that it is required because of future development.
- Applicable boundary conditions
  - The possibility of excluding issues from testing
  - How it is accounted for operational requirements during testing and safety measures
- Definitions of applicable performance criteria
  - The distinction between performance criterion A and performance criterion B (see Table 8.5)
  - Description of safety-related performance criteria and how they are verified. Possible hang-up of safety circuitry might not become apparent during testing

When complex systems and installations are being tested and measured it is necessary that top-level EMC requirements are "translated" into EMC requirements that apply to modules and submodules. For instance, if an installation is extremely large it can become necessary to define its borders where the antennas are placed. In case antennas are placed inside the installation it is necessary to define additional emission/immunity levels.

Many reasons exist to deviate from generic EMC requirements. For instance, if it is foreseeable that the standards do not cover the EM environment in which the equipment is to operate. In that case it is necessary to deviate, possibly extending the frequency range, modulation method, levels, etc. It is recommended to discuss the contents of the test plan with the responsible persons or agency before the EMC tests.

Table 8.5   Performance criteria

Criterion	Description
A	The system/installation continues operation within the predefined limits of its functional specification.
B	The operation of the system/installation is interfered during the presence of the interfering signal. Degradation of operation is allowed within the pre-defined limits of its functional specification. The system/installation resumes undisturbed operation when interference ceases.
C	The operation of the system/installation is interfered and remains interfered during and after the interference signal is present and absent. Intervention of an operator is required to return to fully functional.

Immunity suggests that the equipment under test is never affected by disturbance of any kind or magnitude. It is for this reason that performance criteria are introduced to classify the severity the way interference manifests itself. These performance criteria are defined by the manufacturer before tests are performed with the use of the criteria of Table 8.5. It is important to consider the "bandwidth" for each criterion. Some minor interference might be acceptable when logging images using low resolution yet it might be fully unacceptable under electrosurgery. The interpretation of the results can therefore only be done by the equipment under test (EUT) company's representative and never by the test engineer.

8.3.2 **EMC test report**

A test report documents the observed results of the tests as prescribed in the EMC test plan. The typical content includes the following:

- A description of the equipment under test
  - Make
  - Type and serial number
  - Description of the functionality
- Revision numbers of
  - Hardware
  - Software
  - Firmware

- Photos of the setup during emission measurements and immunity tests. Special attention includes photos where ESD tests are done and the ports where immunity tests are done.
- An overview of test equipment including their calibration due dates.
- A description (make, type) of the test software and test equipment that is used to control the test equipment.
- Overview of used equipment settings during
  - Emission measurements
    - Used measurement bandwidths
    - Scan time
    - Used detector(s)
    - Frequency
  - Immunity tests
    - Used levels
    - Frequency
    - Dwell time
    - Modulation type (AM, FM, PM, etc.)
- An overview of test results: pass or fail
- Test results
  - Emission: graphs that display measured field strength as function of frequency and the applicable limit
  - Immunity: graph that displays the recorded field strength during immunity tests (e.g., radiated immunity) to show that the source actually was operational
  - Observations
- Who's who
  - On behalf of the test agency
  - On behalf of the manufacturer of the tested equipment
- Date when the measurements started and when they are completed
- Signature of the test agency

The *Competent body* is the person who is able to interpret and validate the technical construction file against the EMC directive. The competent body is consulted when no specific EMC standards are used or is deviated from.

The *Notified body* is a third party that has a governmental accreditation based on knowledge and its independent position with respect to involved parties. The notified body, like the competent body, validates the declaration of conformity. The judgment of the notified body can be mandatory for special type of equipment like railway rolling stock and R&TTE equipment.

8.3.3 Responsibilities

Complex installations and complex systems involve large teams; therefore, it is beneficial to clearly define who is responsible for electromagnetic compliance within the (project) organization. This person reports EMC merits of the product to the officer who signs the declaration of conformity. This function can coincide with the safety officer function and at least independency from other departments and managers other than the CEO or business responsible should be guaranteed to prevent short cuts toward obtaining the declaration of conformity.

In the EMC management plan the relevant stakeholders for EMC have been identified. For keeping them informed a *responsibility assignment matrix (RAM)* is advised. A template is given in Table 8.6. In this template, one responsible is identified who receives support from other disciplines. In very complex organizations this template can be extend to a full RACI chart (PMI 2013):

- R: Responsible
- A: Accountable
- C: Consult
- I: Inform

8.4 System-level testing

In the documentation process test reports play key role. On component level, for cables and enclosures, several test methods have been discussed in Chapters 3 and 6. In addition, in most cases for systems and in some situations for installations system level compliance testing is essential. Since these formal measurements take place at specific test sites or in situ, they are costly in both time and money. Therefore, pre-compliance measurements should be made, which give a fair impression if a system is likely to fail or pass.

8.4.1 Pre-compliance measurements

Statistics show that more than 80% of the equipment that is tested for the first time in an EMC lab, fails. In addition, pre-compliance measurements provide valuable information about

Table 8.6 Responsibility assignment matrix (RAM) template

	Project phase	Account manager	Consultant	Purchasing	Lead engineer	Project engineer	Assembly	Service	EMC engineer	Owner
EMC requirement elicitation	Design	●	■						●	
Coordination	Design				■	●			●	
Defining and validating performance criteria	Design				■					
EMC zoning	Design		●		■	●			●	
Design lightning precautions	Design		●		■	●			●	
EMC mitigation in case of zone deviation	Design				■				●	
Reference design	Design				■	●			●	
Alignment of EMC requirements with purchasing plan	Contract	●	■	■						
Verification on correctness and applicability	Contract	●	■	●						
Supply of components with applicable EMC specifications	Contract			●	■					

(Continued)

Table 8.6 (*Continued*) Responsibility assignment matrix (RAM) template

	Project phase	Account manager	Consultant	Purchasing	Lead engineer	Project engineer	Assembly	Service	EMC engineer	Owner
Deliverance of the system conform to EMC standards and regulations	Construction				■	●	●		●	
Adhering to the installation instructions of the suppliers	Construction				■	●	●		●	
Adhering to company EMC instructions	Construction				■	●	●		●	
Training of personnel	Construction				■		●		●	
Verification of EMC mitigation measures	Construction				■		●		●	
Verification of EMC specifications	Final acceptance		●		■	●		●	●	
Validation of EMC specifications	Final acceptance		●		■				●	
Use in conformance with EMC installation criteria	Service							●	●	■
Housekeeping of the EMC zones	Service							■	●	
Interference mitigation	Service							■	●	

■ Responsible.
● Supporting.

the EM-footprint of the EUT, like specific frequencies that are generated on a printed circuit board (PCB). This speeds up troubleshooting when issues are found during full compliance testing or in situ, where the cost per hour is much higher and the availability of the site may be limited.

8.4.1.1 *Emission measurements up to 300 MHz*   Long cables can be efficient antennas, and radiation caused by common mode currents is the dominant disturbance source for frequencies up to 300 MHz. The loop size is nearly always large, even in a common bonding network (see Section 2.3.1); hence a relatively small common mode current may suffice to exceed legal emission limits.

In pre-compliance testing, the common mode current over the cable is measured with a current probe around the cable or cable bundle. For the worst-case orientation of the cable, this current I_{CM} will give rise to an electric field E_{CM} at a distance r of the cable (Goedbloed 1990):

$$E_{CM} = \frac{Z_0 I_{CM}}{2r} \frac{l}{\lambda} \text{ (V/m)} \tag{8.1}$$

where l (m) is the length of the cable, λ (m) is the wavelength, and Z_0 (Ω) is the characteristic impedance in air (which is $120\pi\ \Omega = 377\ \Omega$). When the cable is resonant ($l = \lambda/2$), Eq. 8.1 reduces to

$$E_{CM} = \frac{Z_0 I_{CM}}{4r} \text{ (V/m)} \tag{8.2}$$

and a relation is found between a current measurement which is much easier to execute than the direct field measurement which is required by the standards. IEC 61000-6-3 and the derived harmonized European standard EN-IEC 61000-6-3 state that the emission level is not to exceed 30 dB μV/m at a measurement distance of 10 m. From Eq. 8.2 it can be derived that the current, I_{CM}, then should not exceed 3.3 μA. As a rule of thumb a slightly higher limit may be used and compliance may be expected as long as the common mode current remains below 5 μA (Goedbloed 1990).

For a long cable, the current varies over its length as has been discussed in Section 3.4.4 on transmission lines. The maximum current has to be determined by moving the probe slowly over the cable while performing a continuous measurement. If available on the measurement equipment, the "Max Hold" function should be enabled.

*8.4.1.2 **Immunity*** In pre-compliance testing, radiated immunity testing can be simulated by injecting on purpose a common mode current over cabling. This is called "bulk current injection" used in MIL-STD-461G (2015) measurements (CS114) and the automotive industry (ISO 11452-4 2005). Since measurement probes cannot withstand high power, special current injection probes have to be used for this purpose. For the relation between the electric field specified in the standards and the injection current the following relation can be used (Goedbloed 1990): 10 mA α 1 V/m.

8.4.2 Full-compliance measurements

Full-compliance measurements are used to provide evidence with which the manufacturer can state that his/her product is in compliance with applicable standards. It is deliberately stated that evidence is provided since the end-responsibility remains with the manufacturer.

The open area test site (OATS) has been the general environment for EMC measurements for many years. Since the frequency spectrum is filled with all sorts of transmitters it is hardly possible to distinguish the EUT from the background noise. Measurement chambers are created to deal with this problem but come at a cost.

*8.4.2.1 **Open area test site*** Measurements are often done in open air when possible. Such a test site is called appropriately *OATS*. A conducting ground plane is used to deal with the uncertainty of the earth's soil reflective behavior. A typical layout of an OATS is shown in Figure 8.3 and shows how the measurement distances affect the dimensions of the OATS. The boundary has to be free from reflecting objects such as buildings, fences, or trees. The ground plane does not necessarily have to be made of sheet metal, corrugated metal, or metal gauze can be used without conflict with the standards' requirements. The standard (CISPR 16-1-4 2007) also gives limits for the unevenness of the ground plane.

The main benefit of an OATS is its relative low cost because it does not require a special facilities. The major drawbacks are as follows:

- *Climate*: Rain will affect measurement results. It is undesirable to subject antennas or other equipment to a harsh environment.

- *Background noise*: The background noise has to remain well below the limit line to which is tested. It is very likely

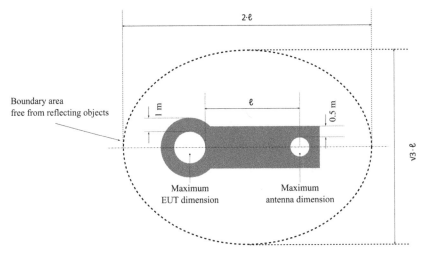

ℓ represents measurement distance

FIGURE 8.3   A typical OATS layout.

that local broadcast stations dominate certain frequency bands that might mask the EUT's emission.

- *Immunity tests*: An unlimited amount of generated noise is prohibited by laws for the same electromagnetic compatibility risks as the EUT that has to undergo these tests. It has to be anticipated upon either getting allowance to transmit at a certain level in a certain bandwidth.

8.4.2.2 Measurement chambers   Measurement chambers are used to create a controlled electromagnetic environment. For mimicking the free space of an OATS, the walls are covered with material that absorb the RF waves. When "illuminated", these absorbers convert the RF energy into heat. Modern versions combine ferrite tiles in combination with triangular shaped foam absorbers for optimizing performance, dimensions, and cost.

The *semi-anechoic room* has a conducting, hence reflecting, ground plane. During emission test the measurement antenna receives both direct and indirect waves (as in an OATS). See Figure 8.4 for an illustration of direct and indirect waves. The *full-anechoic room* uses an absorbing floor and the measurement antenna will only detect the direct wave, which makes analysis easier.

Figure 8.5 shows an example of a measurement enclosure that has two-shielded instrumentation rooms in addition to the main

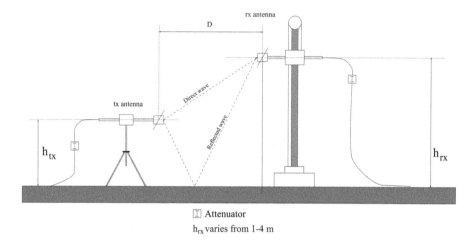

□ Attenuator

h_{rx} varies from 1-4 m

FIGURE 8.4 The EUT that is represented by the transmit antenna "t_x" that emits the direct wave (full- and semi-anechoic) as well as the reflected wave (semi-anechoic) which are registered by the receiving antenna "r_x." The height "h_{rx}" of the receiving antenna is varied from 1 to 4 m.

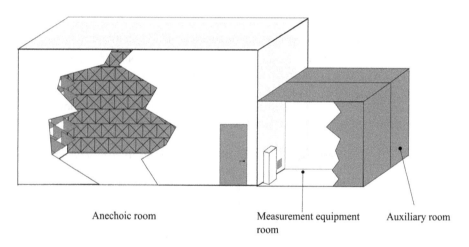

Anechoic room   Measurement equipment room   Auxiliary room

FIGURE 8.5 Example of measurement enclosures.

test chamber. An instrumentation room can be desirable in case of testing very large systems (or *equipment under test [EUT]*) that are installed in different EM environments, as example radar equipment for naval vessels. The main antenna is installed in the exposed environment, whereas the processing equipment that is installed inside the ship in a protected environment. The exposed environment requires immunity to a field strength up to 200 V/m (or more when anticipated) as per MIL-STD 461G (2015).

The protected, or below-decks equipment is to be immune to an EM field strength up to 10 V/m for metallic ships. Splitting the equipment and installing them in separated shielded enclosures facilitates measurement of only the desired subsystem while preventing re-radiation of environmental noise that exists outside the enclosure. This also applies to logging devices that monitor the behavior of the EUT during susceptibility tests.

Measurement equipment like measurement receivers might be installed inside a separate shielded instrumentation room. The EMC cabinet as discussed in Chapter 7 might provide a suitable and more cost-effective alternative.

Cable routing inside the anechoic room requires attention. In case cables are not to be subjected to immunity tests, precautions have to be taken like using cable trays or tubes that are extensively covered in preceding chapters of this book. Note that antenna cables of the measurement facility are among the affected cables due to their transfer impedance. Alternatively battery fed equipment with fiber optics may be used.

EUT-monitoring equipment like cameras (vision) and/or microphone (audio) is also present in these facilities. Safety is a concern, in particular in semi-anechoic enclosures due to the sheet-metal conducting ground plane. An insulating top layer of the ground plane as well as proper tooling and instruction will help improving safety. A trip cord along the inside of the shielded room to activate the safety alarm is recommended as using the wireless phone inside a measurement chamber should not be able to connect to the outside world. In all anechoic room fire prevention measures must be taken, due to the flammability of the absorbers.

8.4.2.3 In situ measurements   Some EUTs, in particular large systems and installations cannot be taken into a measurement facility for all obvious reasons. In some cases, measurements are done *in situ* on the side of the EUT or even inside tunnels, shelters, or naval vessels. In these situations, the measurement uncertainty is high and it is impossible to rule out the effect of the environment:

- *Reflections*: When a measurement antenna is placed inside a metal structure, reflections will affect the results that will differ according to the physical location of the antenna.

- *Absence of a reflecting ground plane*: This may be solved using sheet metal plates, but needs upfront consideration.

- *Measurement distance*: Different measurement distances are sometimes used or the measurement distance is difficult to quantify when placed inside the EUT.

- *Measurement cable routing*: Measurement cable routing affects the reproducibility of the measurements.

- *Background noise*: It is likely that the residual background noise is too high, consequently masking the noise of the EUT. Excessive background noise might also cause overloading an analyzer front end or driving it into saturation.

- *ESD*: For ESD it is not always obvious how to create indirect discharges. As the air humidity plays an important role in these tests, it might be out of range when doing ESD tests in situ.

These measurement errors will occur, but do not disqualify in situ measurements. Sometimes, it is the only manner to gather proof in support of the assumption of conformity which has to be included in the Technical Construction File (TCF). Interpretation of the measurement results will take expert knowledge and has to be documented as well.

8.4.2.4 **Preparations before testing** *Full-compliance testing* has to be initiated when pre-compliance measurements show that the EMC risk has been reduced to a sufficiently low level. This starts with an inspection visit to the *OATS, measurement room*, or in situ site, where the tests will take place. Important points to be considered during inspection are listed in Table 8.7. During the test days it is better not to be dependent on tools, components, or auxiliary equipment of the test laboratory. A packing list is proposed in Table 8.8.

8.4.2.5 **Problem solving** In case the installation or system fails during first tests, the test day may be saved if some quick fixes can be found to solve the noncompliance. First of all check the following:

- Is there visual damage (e.g., due to transport)? Could the damage be relevant?

  A scratch on the paint is unlikely to influence the measurement results, while if a connector shell is broken it could very well matter. In case of relevant damage, the system has to be repaired, such that it meets its original design again.

- Are there any loose connectors?

  When everybody is in a hurry to start testing it can happen that a connector is forgotten, or not well-tightened or it came loose during transport vibration (which then should be noted).

- Are the connections not oxidized (include those of the measurement equipment?

  In particular, in OATS or in outside areas in site, measurement connectors may have become oxidized. In this case they need to be replaced. If the EUT has been stored for a prolonged time, it also should be inspected for oxidization.

- Is the setup conform the measurement standard?

  Not all tests are alike. Make sure that the setup matches the test at hand.

Table 8.7 Checklist for test site inspection

Compressed air	Some equipment require compressed air. Is it available and is the pressure within spec?
Cooling liquid	If required: Is a cooling circuit available? Which coolant is used?
Flow of coolant	Is the pressure within spec? If too high, can it be adjusted according to needs?
Coupling pieces	Are connectors and coupling pieces available to connect the compressed air or the cooling circuit to the equipment under test?
Energy supply	Is the supply voltage ok? Is the power enough? Mains frequency? Phases?
Mains plugs	Will the mains plugs fit? For example, a three-phase 32 A plug would not fit a 64 A socket.
Plastic foil	Some test labs might require foil to put the equipment on to avoid damage to the site when there is a coolant leak. Make sure to bring foil if needed and consider earth connections.
Maximum load of turntable	Does the equipment under test exceed the maximum load of the turntable?
Additional cable length	The turntable will rotate in order to find the worst-case angle, polarization, and antenna height. This will require additional cable length.
Acceleration/deceleration of the turntable	The acceleration/deceleration might be too sudden for equipment that has the majority of the weight in top. The equipment might fall.
Additional equipment	Does the equipment under test need additional or auxiliary equipment for functioning? If so, where does it have to be placed if excluded from the tests?
Obstacles	Are there any obstacles that have to be negotiated when transporting the equipment into the measurement chamber?
Physical location of the measurements	Will all tests and measurements be done on one physical location or is it necessary to move, e.g., when doing immunity measurements.

(Continued)

Table 8.7 (Continued)   Checklist for test site inspection

Performance criteria	Are the performance criteria clearly defined and how they are to be interpreted? This can never be left to the measurement technician.
Door width	Are entrance doors wide enough and high enough?
Accessibility of the test premises	How accessible is the test lab itself?
Emergency stop	Some test labs test their emergency stop buttons every once in a while. If your system is required to run full time because it uses ultra-high vacuum, it is wise to postpone testing the emergency circuitry.
Tools and equipment	Are tools and equipment available at the test location?
Startup time	If the equipment under test needs startup time before it is up and running, this time is to be taken into consideration.
Possibility to leave the test equipment powered	Some companies do not allow the test equipment to remain powered after working hours. This has to be discussed and agreed upon to avoid nasty surprises that might bring the tests to a halt.
Operator	An experienced operator is mandatory. Sometimes it is required to manipulate the software to be able to pinpoint possible noise sources, without disabling too much.
Development organization	Development staff should be aware of the tests and the possibility that their assistance is required on-site.
Mechanics	Are mechanics available on-site to assist?
Fork lift	A fork lift might be needed.
Badges/accessibility	Are badges necessary to gain access to the test premises?
Packaging	Is the equipment under test thoroughly packed? Some test labs might refuse equipment if it is inadequately packed.
Canteen	Very important: food and drinks. What to expect? What to bring?
Working hours	What are the working hours of the test lab?
After working hours	Is it possible to work after official working time? What is there to know?
Storage	Where can the tested equipment be stored and how long? This avoids keeping the test chamber unnecessarily occupied.
Logging possibility	It is necessary to discuss the possibility of data or image logging before immunity tests start. This logging has to be done by the equipment itself and therefore is the responsibility of the manufacturer of the equipment under test.

Table 8.8 Packing list for tools and auxiliary equipment to take to the test site

Adaptors	Adaptors to connect SMA, N-type, and BNC connectors
Coax cable	Measurement cable for diagnostics like RG58 or RG223
Ferrites	You cannot bring enough of them. Bring a wide variety to locate coupling paths
Shielding tape	To close seams and, if needed, to create a temporary ground plane for PCBs
Aluminum foil	Also possibly needed in huge quantities. For example, to create additional shielding or to cover cables
Mains filter	Handy when conducted emissions measurements or conducted immunity tests are done
Tools	It is best to bring your own tools. Maybe some tools are present at the test lab but it is very likely that the collection is limited to hardly ever used screwdrivers
Hot air blower	Some equipment detune slightly when subjected to heat
Extension cord	Some measurement equipment needs a lot of time to startup which can be too time consuming when debugging. Besides, it is not very efficient to repeat entering the right settings
Capacitors	Can be useful to quickly add some capacitors to observe the effect
Soldering iron	Soldering irons are important. There are versions that operate on gas or on batteries. A gas lighter will also do when working at great heights
Pocket knife	To strip wires
Ty-raps	For temporary fixation
Tape	Paper-based tape or light tape is handy to, e.g., mark the points for ESD tests
P-clips	P-clips are useful when wanting to connect the cable braid to chassis the easy way
Marker	To write on tape or whatever
Camera	Make numerous photos and write down what the photo tells. Make notes: it is impossible or hardly possible to tell what happened
Batteries	Spare batteries or accu-pack for the soldering iron, camera, etc.
Attenuators	To protect the front end of the spectrum analyzer
Storage media	To save test results and data that is collected during debugging
Multimeter	For functional checks
Megger	It is wise to check the conductivity of connectors and connector shells to chassis prior to measurements

If the above issues have not occurred or have been fixed without resolving the noncompliance, and the problem cannot readily be identified it may be beneficial to revisit the risk overview and investigate which residual risk may have materialized. In addition, the source–victim matrix (Section 1.1.1) and zone compatibility matrix (Section 1.3.3) may provide useful information.

8.4.2.6 Quick fixes   In Section 8.4.1, it was shown that even small common mode currents (below 5 μA) can give rise to exceeding the emission limits. Culprit cables may be found by measuring the common mode current over cable bundles and then trying to identify the individual cable or cables. A simple, home-made, current probe is shown in Figure 8.6. During measurement one has to take in to account the frequency response of such a probe. As discussed in Section 6.3.1, *ferrites* can be used to suppress and reroute common mode currents and are easy to retrofit during test, in particular snap-on types. Care however must be taken, that the final test results are only valid for the system or installation with the ferrites present, which must be documented and implemented.

In case common mode currents are not the culprit, but direct leakage of fields due to seems or holes (see Section 7.2.4), conductive *shielding tape* can be used to identify if this indeed is the root cause. For optimal result a tape with low resistance

(a)

(b)

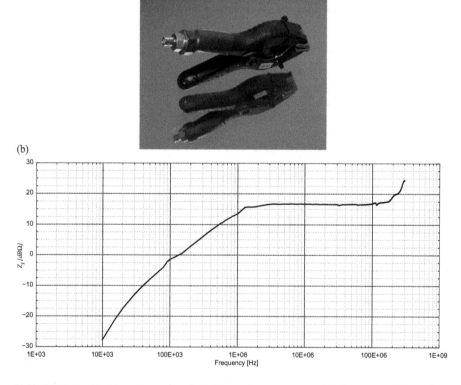

FIGURE 8.6   (a) Current probe for diagnostic purposes; (b) frequency-dependent current–voltage transfer curve (Z_{probe} [dB Ω]): I_{CM} [dB μA] = $U_{measured}$ [dB μV] − Z_{probe} [dB Ω].

and a conductive glue should be selected. Figure 8.7 shows a measurement set-up, with the results from several samples listed in Table 8.9. The attenuation of these samples, measured in a double TEM cell (Wilson and Ma 1985), is shown in Figure 8.8. Local leakage of electromagnetic fields can be identified with *near-field probes* (sometimes called *sniffer probes or coils*). These are small electric or magnetic dipoles as shown in Figure 8.9. The largest magnetic field probe is the most sensitive one because its loop area is the largest. It is useful to globally localize the source. Switching to the smaller ones will eventually pinpoint the source and its orientation since the coupling can be varied by changing the orientation of the loop itself. The electric field probe can be convenient for identifying electric fields. In case of need near field probes can be quickly made from a coaxial, preferably semirigid, cable (see Figure 8.9b).

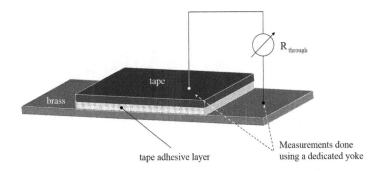

FIGURE 8.7 Setup for measuring the through-resistance of conductive shielding tape.

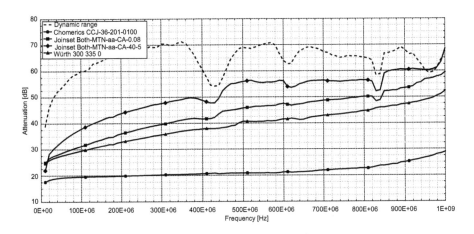

FIGURE 8.8 Measured RF attenuation of conductive shielding.

FIGURE 8.9 Near-field probes for searching fields leaking through seams and holes: (a) commercial set; (b) home-made.

Table 8.9 Test results of conductive shielding tapes

Brand	Type	Width (mm)	$R_{through}$ (mΩ) (see Figure 8.7)	R_{square} (mΩ/cm^2)	Material	Surface structure
3M	1245	19	21.5		Copper	Waffle
Chomerics	CCJ-36-201-0100	25	576.5	1.6	Aluminum	Smooth
Würth	301 322 0	20	811.5	1000	Aluminum	Smooth
Joinset	Both-MTH	40	59.75	7	Aluminum	Smooth

8.5 Subsystem testing

Many equipment cabinets consist of multiple power supplies or computerized equipment. Each and every component will contribute to the total emission of that cabinet. One way to anticipate on the total emission level is to integrate components that meet the emission levels of residential equipment in case the cabinet has to meet industrial levels. This implies that each component's emission level is 6 dB lower than the total emission level. The susceptibility threshold level will also be lower compared to the industrial limit which might require extra shielding by the cabinet and/or filtering.

The general rule of thumb for stochastic sources is that the total emission level will be increased by $10 \log_{10} N$, where N represents the number of similar devices. Where distance is a useful companion reducing radiated phenomena, this is only partially true for conducted interference. When multiple components are fed in parallel from the same power supply, conducted interference can occur.

In case a system is operated with a synchronized clock, care must be taken because disturbances may increase linearly with the number (N) of devices.

CHAPTER NINE

Troubleshooting

The modern approach to achieve electromagnetic compatibility is based on a risk management approach. Risks are identified early in the design and are addressed as soon as possible. After successful full-compliance testing conformance is assumed and risks should be negligible. Nevertheless, incidentally unexpected electromagnetic interference may occur and then troubleshooting is required. A structured approach will help to find the root cause in an efficient manner, i.e., the effort required increases per step:

1. Have changes been made in the installation or system?

    This question typically can be answered without visiting the site of installation. An assessment can be made if the reported interference may be related to the change.

2. Have changes been made in the environment of the installation or system?

    This question typically can be answered without visiting the site of installation. Examples of changes in the environment are the placement of a power transformer in the basement, or mobile antennas on the roof of a nearby building. An assessment can be made if the reported interference may be related to the change.

3. Has software been updated? What is the version number?

    Software contains bugs. Sometimes they introduce intermittent problems which mimic electromagnetic interference problems. When the software version is known, it can be checked for known issues, either in this specific release, or if they have been solved in a newer release.

4. Perform visual inspection.

    a. Are there any loose connectors?

b. Are any of the (ground) connections oxidized or damaged?

    This not only applies to grounding straps, but also to filter connections to enclosures.

c. Have any pigtails been used?

d. Are cables properly separated?

    Cables connected to sensitive equipment should not be routed next to emitting power cables, like variable frequency drive cabling.

5. Perform functional variation.

    Changing the mode of operation of the mal-performing equipment may give indications toward the disturbance cause. Examples are as follows:

a. When an external magnetic field interferes with an electron microscope, the amount of image distortion will depend on the acceleration voltage.

b. The (audible) output noise of an audio amplifier with a shorted input may remain constant when the volume is increased or it may alter.

c. Pulse width-modulated power sources are notorious for their noise production. Changing the output power might change the nature of the observed interference.

6. Verify the input voltages of the mal-performing equipment. Is the interference frequency the mains fundamental or a harmonic? Note that in harsh environments reliable voltage measurements are difficult and skipping to step 7 may be preferred.

7. Perform common mode current measurements.

    Start with measuring common mode currents over cable bundles or grounding structures (see Figure 8.6). When the threshold of 5 µA is exceeded, try to determine which cable(s) in the bundle are the culprit.

8. Perform near-field measurements.

    When common mode currents indicate problems with filters, electromagnetic field leakage between the filter and housing may be confirmed with near-field probes (see Figure 8.9). Also when connections between different parts of an enclosure are marked as suspect during visual inspection, near-field probes will be able to indicate if indeed field is emitted from them.

9. In case problems occur intermittently or only during lightning strikes, inject current into the system and repeat steps 5 and 6 (synchronized with the injection).

References

Books, Journals, and Proceedings

ABB. 2010. *XLPE land cable systems—User's guide*. Rev. 5.

Anderson, R.W. 1967. S-Parameter Techniques for faster, more accurate network design. *Hewlett-Packard Journal*, Vol. 18, No. 6. Also reprinted as HP Application Note 95–1.

Bargboer, G. and A.P.J. van Deursen. 2010. A case study on ligthning protection, current injection measurements, and model. *IEEE Transactions on EMC*, Vol. 52, No.3, pp. 684–690.

Beltran, D., K. Armstrong, A. Charoy and J. Hourtoule. 2011. The ITER earthing topology: Mesh-Common Bonding Network. *IEEE/NPSS 24th Symposium on Fusion Engineering (SOFE)*. Chicago.

Benson, F.A., P.A. Cudd, J.M. Tealby. 1992. Leakage from coaxial cables. *Science, Measurement and Technology, IEEE Proceedings A*, Vol. 139, No. 6, pp. 285–303.

Bridges, J.E. 1988. An update on the circuit approach to calculate shielding effectiveness. *IEEE Transactions on EMC*, Vol. 30, No. 3, pp. 211–221.

Broyde, F., E. Clavelier, D. Givord, P. Vallet. 1993. Discussion of the relevance of transfer admittance and some through elastance measurement results, *IEEE Transactions on EMC*, Vol. 35, No. 11, pp. 417–422.

Butterweck, H.J. 1979. Elektrische netwerken. In: *Het spectrum*. Utrecht: Dutch.

Campione S., L.I. Basilio, L.K. Warne, W.L. Langston. 2016. Transmission-line modeling of shielding effectiveness of multiple shielded cables with arbitrary terminations. *URSI International Symposium on Electromagnetic Theory*.

Chang, H.T. 1980. Protection of buried cable from direct lightning strike. *IEEE Transactions of EMC*, Vol. 22, No. 3, pp. 157–160.

Deursen, A.P.J., F.B.M. van Horck, M.J.A.M. van Helvoort, P.C.T. van der Laan. 1995. Induced currents and voltages in secondary cables connected to Gas Insulated Switchgear. 9ᵗʰ *International Symposium on High Voltage Engineering*, Graz, Austria.

Deursen, A.P.J., M.J.A.M. van Helvoort, F.B.M. van Horck, J. van der Merwe, P.C.T. van der Laan. 2000. Protection of cables by grounding structures. *IEEE Conference on EMC*. Washington, DC.

Degauque, P., J. Hamelin. 1993. *Electromagnetic compatibility*. New York: Wiley.

Fenical, G. 2003. Rule-of-thumb for calculating aperture size. *Laird Tech Notes #154*.

Ferber, R.R., F.J. Young. 1970. Enhancement of EMP shielding by ferromagnetic saturation, *IEEE Transactions on Nuclear Science*, Vol. NS-17, pp. 354–359.

Forsberg, K., H. Mooz. 1991. The relationship of system engineering to the project cycle. *Proceedings of the First Annual Symposium of National Council on System Engineering*.

Getzlaff, M. 2008. *Fundamentals of magnetism*. Berlin: Springer-Verlag.

Goedbloed, J. 1984. *Electro-magnetische compatibiliteit (EMC)*. Gravenhage: PATO.

Goedbloed, J. 1990. *Electromagnetic compatibility*. South Holland: Kluwer.

Grafe, H., J. Loose, H. Kühn, S. Strobach. 1967. *Grundlagen der Elektrotechnik*. Berlin: VEB Verlag Technik.

Hagn G.H., E. Lyon. 1994. The first antenna and wireless telegraph, personal communications system (PCS), and PCS symposium in Virginia. In: Rappaport T.S., Woerner B.D., Reed J.H. (eds) *Wireless personal communications*. The Springer International Series in Engineering and Computer Science, Vol. 262. Boston, MA: Springer.

Harberts, D.W., M.J.A.M. van Helvoort. 2013. Shielding requirements of a 3 T MRI examination room to limit radiated emission. *International Symposium on Electromagnetic Compatibility (EMC Europe 2013)*. Belgium: Brugge.

Harvey, A.F. 1963. Microwave Engineering, Academic Press, New York.

Heaviside, O. 1892. *Electrical papers*. New York: Macmillan.

van Helvoort, M.l.A.M., P.C.T. van der Laan. 1994. EMC demonstrations on cabling and wiring. *ISEMC'94*. Saõ Paulo, Brazil.

van Helvoort, M.J.A.M., A.P.J. van Deursen, P.C.T. van der Laan. 1995. The transfer impedance of cables wih a nearby return conductor and a non-central inner conductor. *IEEE Transactions on EMC*, Vol. EMC-37, No. 2, pp. 301–306.

Helvoort, M.J.A.M. 1995. *Grounding structures for the EMC protection of cabling and wiring*. PhD Thesis. Eindhoven University of Technology, the Netherlands.

van Helvoort, M.J.A.M. 2017. DC currents may induce disturbing voltages in other circuits. *LinkedIn*. www.linkedin.com/pulse/dc-currents-may-induce-disturbing-voltages-other-mark-van-helvoort.

van Houten, M.A. 1990. *Electromagnetic compatibility in high-voltage engineering.* PhD Thesis. Eindhoven University of Technology, the Netherlands.

Kaden, H. 1950/2006. *Die elektromagnetische Schirmung in der Fernmeld- und Hochfrequenztechnik.* Berlin: Springer-Verlag.

Kapora, S., E. Laermans, A.P.J. van Deursen. 2010. Protection of cables by open metal conduits. *IEEE Transactions on Electromagnetic Compatibility.* Vol 52, No 4.

King, L.V. 1933. XXI. Electromagnetic shielding at radio frequencies. *Philosophical Magazine Series 7,* Vol. 15, No. 97, pp. 201–223.

Küpfmüller, K. 1973. *Einführung in die theoretischen Elektrotechnik. 10^e auflage.* Berlin: Springer-Verlag.

Kurokawa, K. 1965. Power and waves and the scattering matrix. *IEEE Transactions on Microwave Theory and Techniques,* Vol. MTT-13, No. 2, pp. 194–202.

Li, M., J.L. Drewniak, S. Radu, J. Nuebel, T.H. Hubing, R.E. DuBroff, T.P. van Doren. 2001. An EMI estimate for shielding-enclosure evaluation. *IEEE Transactions on Electromagnetic Compatibility,* Vol. 43, No. 3, pp. 295–304.

Lindenblad, N.E. 1936. Short wave communication systems. *US Patent.* US2131108.

Luker, J.R. 1998. Selecting magnetic shielding metals—High-permeability shielding materials prevent interference from driving sensitive circuits crazy, *Machine Design,* May 21.

Maciel, D. 1993. *Etude et modélisation de resiques électromagnétiques supportés par des cables de transmission d'informations contenus dans des Chemins métalliques insallés sur des sites industriels.* In French. PhD Thesis. Lille University of Science and Technology, France.

Martin, A.R. 1982. An introduction to surface transfer impedance. *EMC Technology,* Vol. 1, No. 3, pp. 44–52.

Melenhorst, M., M. van Helvoort. 2015. *Grounding of multi-cable transits.* www.alewijnse.com/sites/default/files/bestanden/downloads/grounding_of_multi_cable_transits_for_on-shore_use.pdf.

Merewether, D.E. 1970. Analysis of the shielding characteristic of saturable ferromagnetic cable shields. *IEEE Transactions on EMC,* Vol. EMC-12, pp. 134–137.

Merewether, D.E. 1970. Design of shielded cables using saturable ferromagnetic materials. *IEEE Transactions on EMC,* Vol. EMC-12, pp. 138–141.

Merwe, J.S., H.C. Reader, D.J. Rossouw. 2011. Cable tray connections for electromagnetic interference (EMI) mitigation. *IEEE Transactions on EMC,* Vol. 53, No. 2, pp. 332–338.

Miller, D.A., J.E. Bridges. 1968. Review of circuit approach to calculate shielding effectiveness. *IEEE Transactions on EMC,* Vol. 10, No. 52, pp. 52–62.

Ogunsola, A., A. Mariscotti. 2013. *Electromagnetic compatibilty in railways—Analysis and management.* Berlin, Heidelberg: Springer.

Ott, H.W. 1983. Ground—A part for current flow. *EMC Technology.* January–March: 44–48.

Ott, H.W. 2009. *Electromagnetic compatibility engineering*. Hoboken, NJ: Wiley.

Peterson, H.O. 1945. Antenna construction. *US Patent*. US2478313.

Philippow, E. 1968. *Taschenbuch Elektrotechnik*. Berlin: VEB Verlag Technik.

Phipps, K.O., P.F. Keebler. 2008. Understanding shielding effectiveness of materials and measurements in the near-field and far-field. *Interference Technology*, https://interferencetechnology.com/shielding-effectiveness-near-field-far-field.

PMI. 2015. *Business analysis for practitioners: A practice guide*. Newtown Square, PA: Project Management Institute.

PMI. 2013. *A guide to the project management body of knowledge*. Newton Square, PA: Project Management Institute.

Quednau, G. 2014. *Montage von Kabelschirmen unter EMV-Aspekten* (in German). Huckeswagen: Pflitsch.

Ramo, S., J.R. Whinnery, T. van Duzer. 1984. *Fields and waves in communication electronics*. New York: Wiley.

Robbins, T. 2017. *Valve amplifier hum*. www.dalmura.com.au.

Ruark, A.E., M.F. Peters. 1926. Helmholtz coils for producing uniform magnetic fields. *Journal of the Optical Society of America*, Vol. 13, No. 2, pp. 205-212.

Schaffner. 1996. CISPR 17 measurements 50 Ω/50 Ω versus 0.1 Ω/100 Ω. *Note 690-264A*.

Schelkunoff, S.A. 1934. The electromagnetic theory of coaxial transmission lines and cylindrical shields. *Bell System Technical Journal*, Vol. 14, pp. 532–579.

Schelkunoff, S.A. 1938. The impedance concept and its application to problems of reflection, refraction, shielding and power absorption. *Bell Systems Technical Journal*, Vol. 17, No. 1, pp. 17–48.

Schelkunoff, S.A. 1943. *Electromagnetic waves*. Princeton, NJ: Van Nostrand.

Sunde, E.D. 1945. Lightning protection of buried toll cable. *Bell Systems Technical Journal*, Vol. 24, pp. 253–300.

Sunde, E.D. 1968. *Earth conduction effects in transmission systems*. New York: Dover.

Teuchert, H., K. Wahl. 1959. *Grundlagen der Elektrotechnik*. Leipzig: Fachbuchverlag Leipzig.

Vance, E.F. 1974. Shielding effectiveness of braided wire shields. *Interaction Note 172*. Menlo Park, CA: Stanford Research Institute.

Vance, E.F. 1975. Shielding effectiveness of braided-wire shields. *IEEE Transaction on EMC*, Vol. EMC-17. No. 2. pp. 71–77.

Waldron, A. 2007. EMC for sound and TV installations. *IET Seminar on New Regulatory Requirements and Techniques for Achieving Electromagnetic Compatibility in Commercial Buildings*, pp. 91–110.

Wilson, P.F., M.T. Ma. 1985. Shielding effectiveness measurements with a dual TEM cell. *IEEE Transactions on Electromagnetic Compatibility*, Vol. EMC-27, No. 3, pp. 137–142.

Worm, S.B., L.L. Kanters. 2004. Shielding effectiveness analysis of DVD recorder enclosures. *International Symposium on Electromagnetic Compatibility (EMC Europe 2004)*. The Netherlands: Eindhoven.

Wouters, P.A.A.F., M.J.A.M. van Helvoort, P.C.T. van der Laan. 2000. Reduction and control of ELF magnetic fields, *Tutorial session, EMC 2000 Brugge Symposium*, Belgium.

Yang, B. et al. "Numerical Study of Lightning-Induced Currents on Buried Cables and Shield Wire Protection Method" (IEEE, April 2012)

Zygology, Galvanic Corrosion Chart. http://www.zygology.com/cms/upload_area/pdf/Zyg-Anodic-Index.pdf.

Standards and Directives

ANSI/IEEE 518–1982. 1996. *IEEE guide for the installation for electrical equipment to minimize electrical noise inputs to controllers from external sources.*

ANSI/SCTE 51. 2012. *Method for determining drop cable braid coverage.* Society of Cable Telecommunications Engineers.

ASTM A 698/A. 2002. *Standard test method for magnetic shield efficiency in attenuating alternating magnetic fields.*

ASTM D 4935—10. *Standard test method for measuring the electromagnetic shielding effectiveness of planar materials.*

CISPR 16-1-4. 2007. *Ancillary equipment—Radiated disturbances and immunity measuring apparatus—Antennas and test sites for radiated disturbance measurements.*

CISPR 17. 2011. *Methods of measurement of the suppression characteristics of passive EMC filtering devices.*

Directive 2006/42/EC. 2016. Directive 2006/42/EC of the European parliament and of the council of 17 May 2006 on machinery, and amending Directive 95/16/EC. *Official Journal of the European Union*, Vol. 49, pp. 24–88.

Directive 2014/30/EU. 2014. Directive 2014/30/EU of the European parliament and of the council of 26 February 2014 on the harmonization of the laws of the Member States relating to electromagnetic compatibility. *Official Journal of the European Union*, Vol. 96, pp. 79–106.

IEC 11801. 2009. *Information technology—Generic cabling for customer premises.*

IEC 60050-161:1990/AMD7. 2017. *International electrotechnical vocabulary—Part 161: Electromagnetic compatibility.*

IEC 60364-1. 2005. *Low-voltage electrical installations—Part 1: Fundamental principles, assessment of general characteristics, definitions.*

IEC 60364-1. 2002. *Low-voltage electrical installations—Part 7: Requirements for special installations—Medical locations.*

IEC 60364-5. 2011. *Low-voltage electrical installations—Part 54: Selection and erection of electrical equipment—Earthing arrangements and protective conductors.*

IEC 60439-1. 2004. *Low-voltage switchgear and control gear assemblies—Part 1: Type-tested and partially type-tested assemblies.*

IEC 60512-23-3. 2000. *Electromechanical components for electronic equipment—Basic testing procedures and measuring methods—Part 23-3: Shielding effectiveness of connectors and accessories*, Ed. 1.0.

IEC 60601-1. 2005. *Medical electrical equipment—Part 1: General requirements for basic safety and essential performance*, Ed. 3.0.

IEC 60601-1-2. 2014. *Medical electrical equipment—Part 1-2: General requirements for basic safety and essential performance—Collateral standard: Electromagnetic disturbances—Requirements and tests.*

IEC TR 61000-2-5. 2017. *Electromagnetic Compatibility (EMC)—Part 2-5: Environment—Description and classification of electromagnetic environments.*

IEC 61000-4-5. 2014. *Electromagnetic Compatibility (EMC)—Part 4-5: Testing and measurement techniques—surge immunity test.*

IEC 61000-5-2. 1997. *Installation and mitigation guidelines—Part 5: Installation and mitigation guidelines—Section 2: Earthing and cabling.*

IEC/TR 61000-5-6. 2002. *Electromagnetic Compatibility (EMC)—Part 5-6: Installation and mitigation guidelines—Mitigation of external EM influences.*

IEC 61000-5-7. 2001. *Electromagnetic Compatibility (EMC)—Part 5-7: Installation and mitigation guidelines—Degrees of protection provided by enclosures against electromagnetic disturbances (EM code).*

IEC 61000–6-1. 2005. *Electromagnetic Compatibility (EMC)—Part 6-1: Generic standards—Immunity standard for residential, commercial and light-industrial environments.*

IEC 61000–6-2. 2005. *Electromagnetic Compatibility (EMC)—Part 6-2: Generic standards—Immunity for industrial environments.*

IEC 61000–6-3. 2010. *Electromagnetic Compatibility (EMC)—Part 6-3: Generic standards—Emission standard for residential, commercial and light-industrial environments.*

IEC 61000–6-4. 2018. *Electromagnetic Compatibility (EMC)—Part 6-4: Generic standards—Emission standard for industrial environments.*

IEC 61508-1. 2010. *Part 1: Functional safety of electrical/electronic/programmable electronic safety-related systems.*

IEC 61537. 2006. *Cable management—Cable tray systems and cable ladder systems.*

IEC 61587. 2013. *Mechanical structures for electronic equipment—Tests for IEC 60917 and IEC 60—Part 3: Electromagnetic shielding performance tests for cabinets and subracks.*

IEC 61726. 2015. *Cable assemblies, cables, connectors and passive microwave components - Screening attenuation measurement by the reverberation chamber method.*

IEC 62153-4-3. 2013. *Metallic communication cable test methods—Part 4-3: Electromagnetic compatibility (EMC)—Surface transfer impedance—Triaxial method*, Ed. 2.0.

IEC 62153-4-6. 2013. *Metallic communication cable test methods—Part 4-6: Electromagnetic compatibility (EMC)—Surface transfer impedance—Line injection method*, Ed. 2.0.

IEC 62153-4-4. 2015. *Metallic communication cable test methods - Part 4-4: Electromagnetic compatibility (EMC) - Test method for measuring of the screening attenuation as up to and above 3 GHz, triaxial method.*

IEC 62153-4-7. 2006. *Metallic communication cable test methods—Part 4-7: Electromagnetic compatibility (EMC)—Test method for measuring the transfer impedance and the screening—Or the coupling attenuation—Tube in tube method.*

IEC 62153-4-9. 2018. *Metallic communication cable test methods - Part 4 - 9: Electromagnetic compatibility (EMC) - Coupling attenuation of screened balanced cables, triaxial method.*

IEC 62305-4. 2012. *Protection against lightning—Part 2: Risk management.*

IEC 62305-4. 2012. *Protection against lightning—Part 4: Electrical and electronic systems with structures*, Ed. 2.0.

IEEE Std. 142. 2007. *IEEE recommended practice for grounding of industrial and commercial power systems.*

IEEE Std. 299. 2006. *IEEE standard method for measuring the effectiveness of electromagnetic shielding enclosures.*

IEEE Std. 299.1. 2013. *IEEE standard method for measuring the shielding effectiveness of enclosures and boxes having all dimensions between 0.1 m and 2 m.*

IEEE Std. 802.3. 2012. *IEEE standards for local and metropolitan networks.*

IEEE Std. 1100. 1999. *IEEE recommended practice for powering and grounding electronic equipment.*

IEEE Std.1302. 2008. *IEEE guide for the electromagnetic characterization of conductive gaskets in the frequency range of DC to 18 GHz.*

ISO/IEC. 1996. *Guide 2.*

ISO 11452-4. 2005. *Road vehicles—Component test methods for electrical disturbances from narrowband radiated electromagnetic energy—Part 4: Bulk current injection (BCI).*

MIL-HDBK-419. 1987. *Grounding, bonding, and shielding for electronic equipments and facilities.* Volume 2 of 2 Applications.

MIL-STD-285. 1965. *Military standard: Attenuation measurements for enclosures, electromagnetic shielding, for electronic test purposes, Method of.*

MIL-STD-461G. 2015. *Department of defense interface standard: Requirements for the control of electromagnetic interference characteristics of subsystems and equipment.*

NPFA 70. 2014. *National electric code.* Quincy, MA: National Fire Protection Association.

Websites

Chalfant, www.chalfantcabletray.com.
IEC-CISPR 16, www.iec.ch/emc/basic_emc/basic_cispr16.htm.
Intertek, www.ieee.li/pdf/essay/guide_to_global_emc_requirements_
2007.pdf.
MCB, www.mcbboek.nl.
Ott, www.hottconsultants.com/regulations.html.

YouTube

van Helvoort, M.J.A.M. 2017. https://youtu.be/ZWnuH0qdly0, elector.tv.

Index